DISCOVERING GENETICS

Cover picture: A model of part of the DNA molecule, with its coded sequence of bases specifying characteristics of a living organism.

Opposite: These tiny zebra fish have developed from eggs that have been treated by heat in such a way that they can be used to provide offspring genetically identical to themselves in subsequent generations.

DISCOVERING
GENETICS

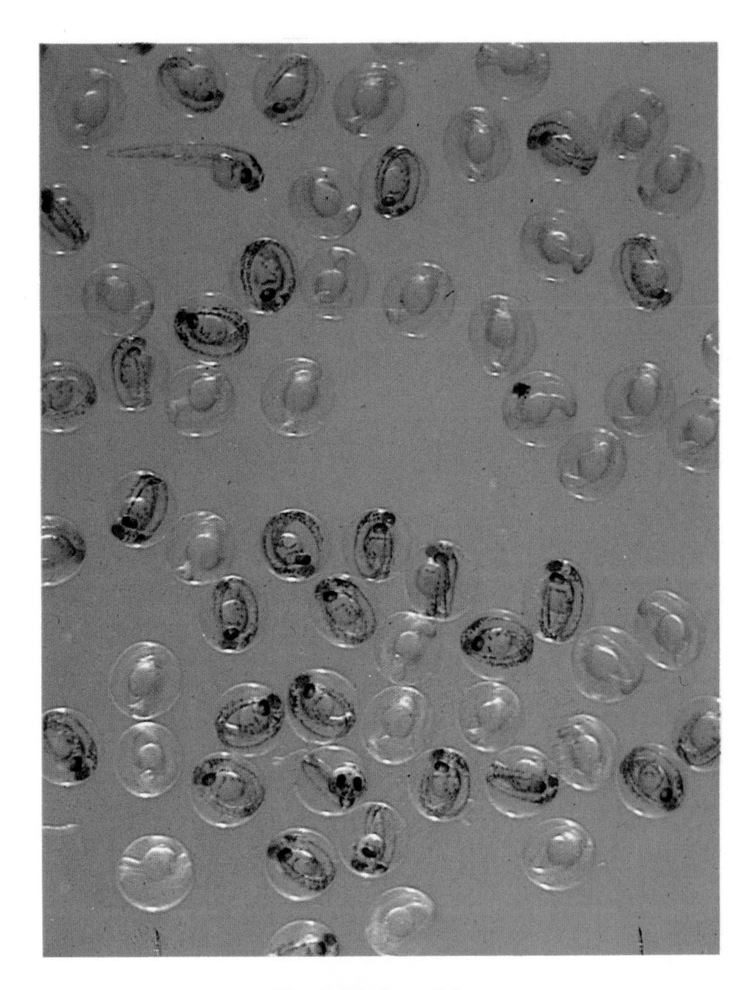

Published by

in association with

The American Museum of Natural History

The consultant

Dr Charles J. Cole is a curator in the Department of Herpetology of the American Museum of Natural History in New York City. He specializes in the biology of amphibians and reptiles, concentrating on their genetics, reproduction and evolution. His current work deals largely with unisexual (all-female) species of lizards. His field and laboratory investigations have taken him throughout most of the United States and Mexico, and to the West Indies, South America and Europe.

The American Museum of Natural History

Stonehenge Press wishes to extend particular thanks to Dr Thomas D. Nicholson, Director of the Museum, and Mr David D. Ryus, Vice President, for their counsel and assistance in creating this volume.

Stonehenge Press Inc.:
Publisher: John Canova
Editor: Ezra Bowen
Deputy Editor: Carolyn Tasker

Created, designed and produced by
Trewin Copplestone Books Ltd, London

Library of Congress Card Number 81-52419
Printed in USA by Rand McNally & Co.
First printing

ISBN 0-86706-010-7
ISBN 0-86706-061-1 (lib. bdg.)
ISBN 0-86706-030-1 (retail ed.)

Set in Monophoto Rockwell Light by
SX Composing Ltd, Rayleigh, Essex, England
Separation by Gilchrist Bros. Ltd, Leeds, England

Contents

The World of Genetics

Archaeologists have discovered that as long as seven thousand years ago, farmers in Central America were improving their crops of corn by sowing seed taken from plants that had developed desirable characteristics. These ancient farmers were exploiting their knowledge of what is now called inheritance. To them as to us, everyday observation revealed that in all living organisms – including human beings – certain things run in families. We expect the color of hair and eyes, the shape of chins and noses, and many other characteristics to be inherited – but how? Thanks to the scientific study of inheritance – the science of genetics – the answer to this fundamental question is now largely known.

Geneticists have shown how biological instructions for particular characteristics are handed down from generation to generation, and have revealed the chemical substance on which these instructions are transmitted. They have also demonstrated that an organism is the product both of what it inherits, and the environment in which it lives – for example, the climate, the quality and quantity of its nutrition, and the conditions inside the egg or seed or womb in which it develops.

Genetics is more than just the study of inheritance in present-day organisms. Assuming that all living things have evolved from the same primitive ancestor, geneticists also ask how inheritance operates in the long term, and how the earth's incredible variety of living organisms has arisen. They also seek to discover how biological instructions have changed, leading to the development of new species, and to uncover the powerful influences that shape the paths of evolution.

But genetics is concerned not only with the how and why of inheritance. It also has a more practical side. It has demonstrated that certain diseases are inherited, helping doctors in their continuing quest to control, and eventually eradicate, disease. Genetics has also made vital contributions to the science of agriculture, increasing the success of animal and plant breeders as they attempt to improve the quality of their livestock and crops.

The latest advances in genetics enable scientists to manipulate directly some of the characteristics an organism inherits. One day it may be possible to create organisms that are precisely suited to our particular needs. Such genetic engineering promises to provide ways of controlling and enriching our environment on a scale undreamed of by our ancient ancestors.

What is Inherited?

Physical characteristics can be inherited over many generations, as in the famous protruding lip and jaw that occurs in the Hapsburg royal family of Austria. The features of members of the Hapsburg family are well recorded and provide scientists with a useful case history for studying the effects of inheritance. Shown from left to right are: Rudolf I (1218–91), Charles V (1500–58), Charles II (1661–1700) and Archduke Albert (1817–95).

It would be surprising if a spaniel gave birth to a terrier, or seeds from a poppy grew into daffodils. If a female Siamese cat produced striped kittens, we would suspect that the tomcat from next door had paid a visit. We also expect that children will look like their parents in some respects.

Everyone must be aware that plants and animals always produce offspring similar to themselves. However, occasional differences do occur. Scientists and breeders have tried to exploit this fact by deliberately breeding plants and animals to suit human purposes: flowers of particular kinds to ornament gardens, cereal crops good for flour, dachshunds to chase badgers, greyhounds and horses for racing, and so on. But which characteristics are inherited and how? Does intelligence run in families? Are certain disabilities and diseases such as deafness, diabetes, cancer and heart trouble inherited? Is athletic ability or musical talent inherited?

To begin to answer these and hundreds of similar questions scientists must clarify what they mean by inheritance. They must distinguish between biological inheritance – what is inherited from one's parents at the moment of conception – and environmental inheritance or acquired characteristics – for example, the mother tongue or native language one speaks, which depends on what is heard during childhood. It is often difficult to decide whether inheritance is biological or environmental. For example, although the appearance of red curly hair in successive generations is clearly an example of biological inheritance, what about an ability to play the piano well? Is this due to a biological mechanism or is it due just to a child's hearing and being encouraged to play piano music? Need this ability be due just to one cause or could it be a result of both biological and environmental factors? Mozart developed his amazing musical talent by the age of four. Was this just because he had practiced hard, or was he born with a special ability?

The precise answers to these questions are elusive. However, what human beings, and all other living organisms, do inherit is a series of tendencies towards certain characteristics. Whether these characteristics actually develop will often depend on the environment in which the organism lives, both before and after birth. Even physical features, such as the shape of eyes or ears, or the color of hair, though

A tiny pony, dwarfed beside a powerful cart-horse, shows that there can be enormous variation within the same species (a group of organisms that can interbreed to produce fertile offspring). Each breed retains its particular characteristics if bred with others of its kind.

This offspring of a mating between a Siamese cat and a tabby cat has inherited the characteristic Siamese buff body, but the head and legs have tabby stripes.

largely biologically inherited, can be modified by the environment. For example, some African children suffering from malnutrition have red hair instead of black, and the color of flowers can depend both on what seeds they grow from, and the soil in which they grow. For example, hydrangeas, which have blue flowers when grown in a clay soil, develop flowers with a reddish tinge when grown in chalky soil. Yet there are strict limits to how far the environment can change biologically inherited tendencies. No amount of environmental juggling is likely to produce a puppy from two cats!

How can a tendency be inherited? Obviously blue eyes or petals or broad-feathered wings, all of which are biologically inherited, are not themselves present in a fertilized egg. Yet at this stage the biological characteristics of the parents have already been passed on to the new organism. The instructions for passing on these characteristics are called genes and their study is known as genetics. The science of genetics sets out to answer such questions as: What is the genetic inheritance? What is it made of? How does it work?

Genes and Mutations

Brothers or sisters are rarely identical with one another, but often resemble their parents in different ways. This is because, in every organism, there are many thousands of different genes determining a multitude of detailed characteristics. As a result, except in the case of identical twins, who develop from the same fertilized egg, the chances of any two children having exactly the same combination of their mother's and father's genes are infinitesimally small.

Genes are able to pass on the same characteristics over many generations. However, on rare occasions new characteristics do appear: a long-stemmed plant that appears with a short stem, or a straight-winged fly with curly wings. These altered characteristics may then be passed on to future generations, although they may not appear in every individual or generation. A change of this sort is due to a mutation, a change in a gene so that its instructions are altered. Both the altered gene and the organism possessing it are said to be mutant. The more common nonmutant form of the gene, and the corresponding organism, are called the "wild-type" – the typical form that occurs in nature.

Mutants are the raw material on which geneticists often depend. By comparing the inheritance of a mutant characteristic with that of its nonmutant counterpart, they can learn much about the hidden mechanism of inheritance. But mutations are rare. For any particular characteristic, a mutation generally occurs only once in several tens or hundreds of thousands of individual organisms. So in their work geneticists generally use organisms that breed quickly or produce large numbers of offspring. Mice are often studied, but much more rapid breeders such as bacteria, molds, and flies are even more useful. Geneticists have also learned ways of obtaining rare mutants. One way is to increase the rate at which mutations occur, for example, by subjecting organisms to certain chemicals or to forms of radiation such as X-rays or ultraviolet light. Another way is to make sure of detecting a mutant when it turns up. Although a fly with curly wings rather than the wild-type straight wings is easy to spot, a blind fly is more difficult to detect. In order to discover blindness geneticists must give flies eye tests, by determining, for example, if they move toward light.

Not all organisms reproduce sexually, but those that do inherit two sets of genes when the egg, which contains genes from the mother, is fertilized by sperm (or in the case of plants, pollen) containing genes from the father. For each characteristic there are two sets of instructions.

If an organism inherits a pair of genes that are both in the wild-type form, or both in the mutant form, there is no conflict. But occasionally one gene is in the wild-type form, and the other is in the mutant

Because it breeds rapidly the fruit, or vinegar, fly is very popular with researchers and it has played an important part in the development of modern genetics. Shown here are a normal fly with red eyes (right) and a mutant form with white eyes (far right).

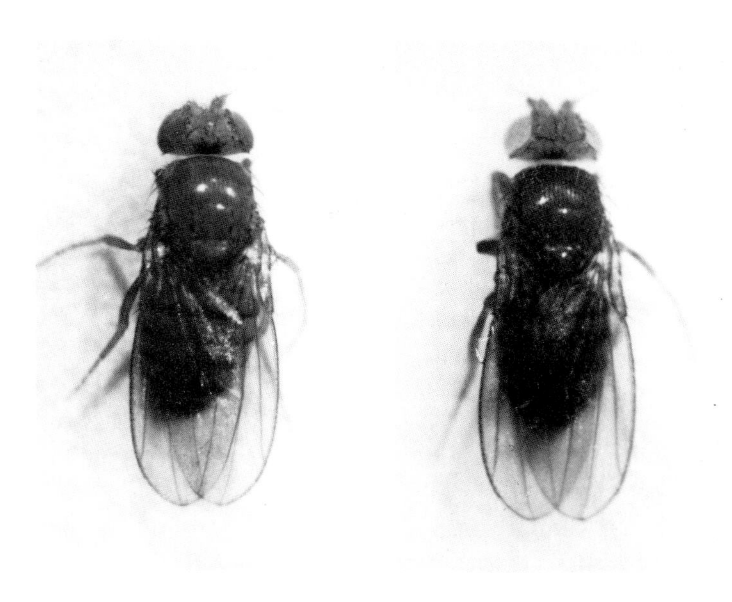

form. For example, a gene from one plant may be an instruction for making long thin petals, while that from the other plant is an instruction for round petals. In such cases there are two possible alternatives. Both instructions may be obeyed, producing an intermediate form (say, oval, medium length petals), or one gene may dominate the other, so that the characteristic only appears in the form which the dominant gene determines.

A gene that dominates is called a dominant gene, and its counterpart is called a recessive gene. So even when a gene is present, it may not influence the appearance of an organism if that gene is inherited along with an alternative, dominant form. In fact, certain genes are found to be always dominant and others always recessive.

The albino deer on the right contrasts sharply with its normal counterpart. The total lack of color results from a mutation that causes a failure in the production of pigment in its skin, hair and eyes.

Cells and Organisms

All living organisms have genes, but if genes are instructions, what form do they take? To discover this requires a search into the structure of living organisms and a look at where genes operate in the microscopic world of the living cell. All living organisms are made up of compartments called cells. Most cells are so tiny that about 100 could be lined up across the head of a pin. The human body consists of billions of cells, but many smaller organisms contain far fewer. Some, for example amoebas and bacteria, are just single cells.

Although most individual cells are so small they can be seen only under a powerful microscope, there are a few exceptions. An egg, even if it is as big as an ostrich's ($6\frac{1}{2}$ inches long), is just one cell. Some human nerve cells have fibers up to three feet long.

Complex organisms, such as human beings, contain many different types of cell, and each type has a specific function. Some of the cells in the blood, for example, can change shape so as to surround and gobble up invaders. Some types of nerve cell have many branches, which enable these nerve cells to "converse" with neighboring cells and to help form a complicated communications network like a telephone exchange. There are cells in the nose that respond to particular chemicals by producing electrical signals that the brain interprets as smell. There are also cells in the eyes sensitive to light. Some cells in plant leaves control the opening and closing of tiny holes to permit such gases as carbon dioxide and oxygen to flow in and out of the leaves. An animal or plant is a community of cells of many types that cooperate to ensure the smooth functioning of the organism.

Despite this variety of functions, there are certain features shared by many cells. A typical animal cell is enveloped by a thin barrier, called the cell membrane, that helps determine which substances are allowed in and out of the cell. In each cell is a spherical structure, itself surrounded by a double membrane barrier. This is the cell nucleus, which is packed with material vital for the cell's survival and reproduction. The substance surrounding the nucleus, known as cytoplasm, contains several additional structures: mitochondria (singular is mitochondrion), which the cell uses to convert foodstuffs to provide energy and the many roundish objects called ribosomes where proteins are manufactured. In cells that make a lot of protein, for example, liver cells, ribosomes form up to twenty-five per cent of the cell's dry weight. Ribosomes are often attached to a complex network of membranes that extend through the cytoplasm of the cell. These membranes form channels through which substances pass around, and occasionally out of, the cell.

Plant cells are similar to animal cells but have a tough outer cell wall, made of cellulose, surrounding the cell membrane. Many cells in green plants also contain up to fifty chloroplasts, sausage-shaped

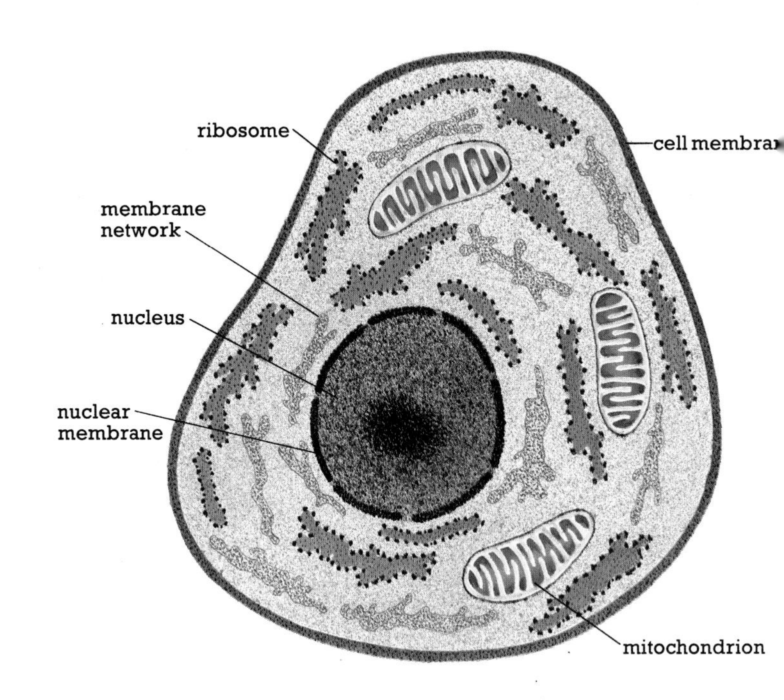

This schematic section through a typical animal cell shows its single nucleus, three of its many mitochondria, and the ribosomes that appear as tiny dots densely studded along the cell's network of internal membranes.

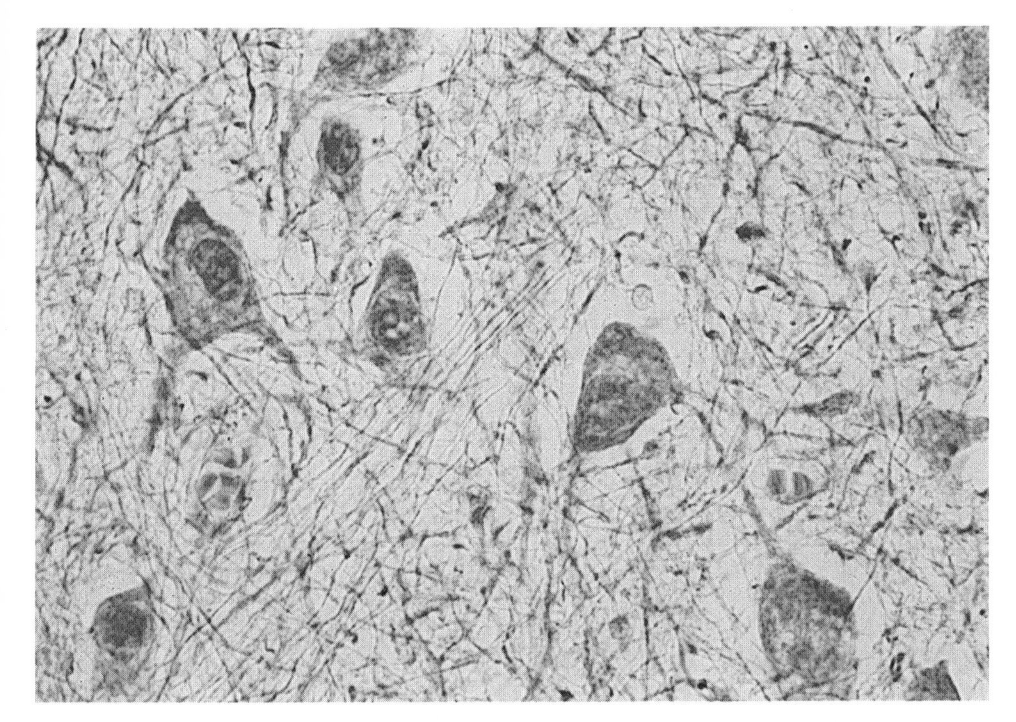

Nerve cells from the brain of a human shown here magnified 100 times, have a central body from which long fibers extend, forming pathways for the transfer of signals between cells.

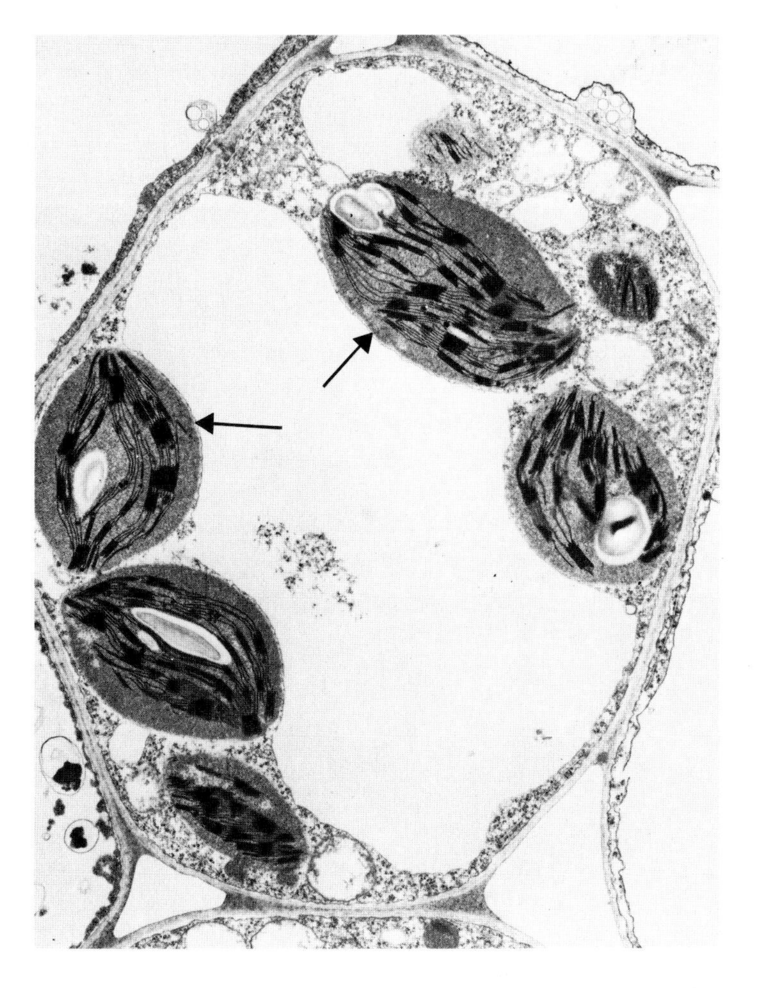

Plants cells, such as the one at left, contain chloroplasts (indicated by arrows) where carbon dioxide absorbed from the air is used to produce sugars.

structures within which are many layers of membrane. Chloroplasts act as tiny solar batteries, enabling plants to trap energy from sunlight. The energy is used to produce sugars from carbon dioxide, which is absorbed by the plants' leaves from the surrounding air. Chloroplasts contain chlorophyll, a green substance that gives many plants their familiar color.

The single cells of bacteria are much less sophisticated in structure than those in the organisms described above. They lack the internal membranes found in animal and plant cells. There is no clearly visible nucleus and they do not have mitochondria. Even so, bacteria can carry out some of the same vital functions as more complex cells, such as the production of hereditary material.

The Chemistry of the Cell

The particular characteristics of any organism largely depend on the types and number of its cells and the way they act together as a community. The specific characteristics of the individual cell – whether it is of a single-celled organism, or just one cell in a multicellular organism – depend in turn on the structures and chemical substances that the cell contains. These structures and substances ultimately depend on the cell's surroundings and its inherited instructions, its genes. How genes perform their remarkable function is revealed in the day-to-day running of a cell.

Cells are highly sophisticated chemical factories. After food is taken in and digested by an organism, the products of digestion are taken up by a variety of cells. In these cells the products are converted to all the thousands of different chemical substances and structures that go to make up the cells and the organism as a whole. Some of the chemical substances inside the cells are composed of fairly small molecules, and include sugars such as glucose, or fatty acids similar to those present in vegetable cooking oil or margarine.

Cells break down small molecules to release energy for the many functions they have to perform. However, cells can also join small molecules together to produce much larger molecules, such as polysaccharides (of which starch and cellulose are examples) and proteins. These can be hundreds or even thousands of times the size of a small molecule such as glucose. Protein molecules often play an important part in the structure of a cell, helping to determine its characteristic appearance. About half of the weight of the mitochondria and cell membrane is made up of such structural proteins. Polysaccharides also have a structured role: one particular type, cellulose, gives cell walls of plants their toughness and rigidity.

A cell's ability to produce thousands of different chemical substances rapidly and precisely depends on a range of proteins called enzymes, whose function is to speed up chemical reactions. Each type of enzyme speeds up just one of the thousands of different reactions that go on in a cell. So the chemical makeup, and the particular characteristics of any cell, are influenced by what enzymes it contains.

Enzymes, like all other proteins, are large molecules that consist of one or more chains of small molecules called amino acids. There are twenty different types of amino acids in these chains. Typically, a chain contains several hundred amino acids joined together like beads in a necklace. The number of possible different sequences of amino acids (that is, the number of possible different types of chain) is

Chemicals in a cell are generally produced by a sequence of reactions. For example, in the hypothetical sequence below, leading from substance A to the production of substance D, there are three reactions, each speeded up by a specific enzyme. In this schematic representation the chemical structures of the three enzymes, E1, E2, and E3, are each determined by a particular gene. In this way, genes can control what substances – for example, sugars – a cell produces.

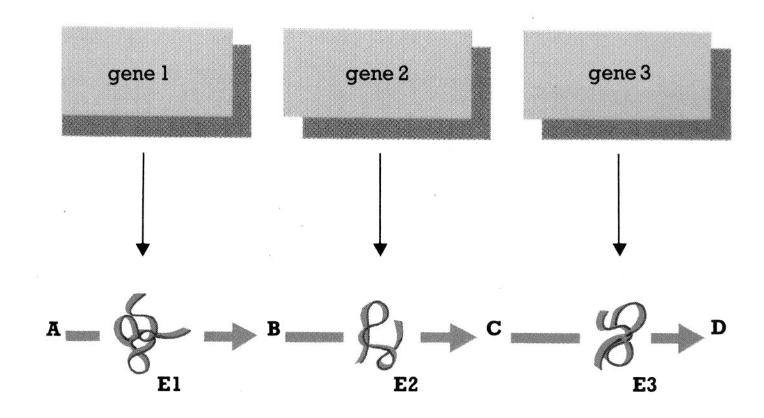

enormous. In some proteins with more than one chain, all of these chains are identical; in other proteins there is more than one type of chain.

Such chains are tightly folded up to form complex three-dimensional shapes. Each type of enzyme has a characteristic shape that enables specific substances to attach themselves to it and become chemically converted. The shape a chain folds up into is in turn determined by the precise sequence of amino acids. This is because the properties of the individual types of amino acid differ. Some carry tiny electric charges that enable them to sit comfortably in the watery environment of the cell; others are oily and so are repelled by water. The chain of acids folds up with the electrically charged acids exposed on the outside and the oil acids hidden inside.

All proteins are themselves manufactured by the cell, and in order to ensure that each type of protein contains its amino acids in the correct sequence the cell must have instructions. These instructions are the genes. For each type of amino acid chain there is one specific gene. So in proteins with more than one type of amino acid chain, a different gene dictates the sequence of each type of chain.

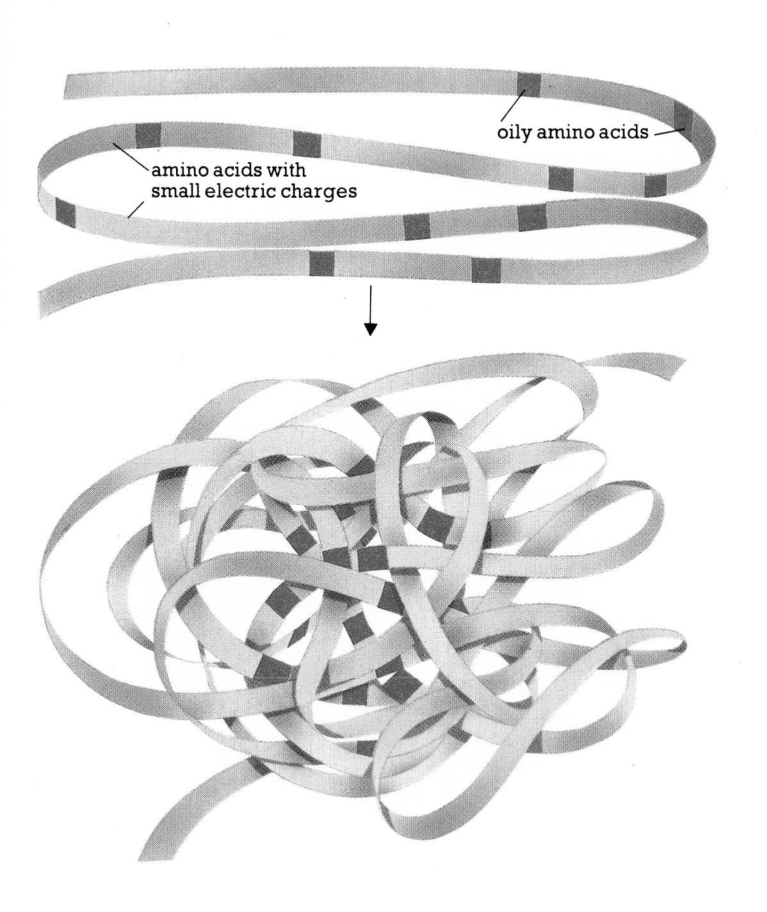

Once it has been created as an extended chain of amino acids (top), an enzyme folds up in a distinctive way. As it does so, the amino acids that are oily tend to seek the inside, while others that carry electric charges (shown in yellow), lie on the outside, in contact with the watery environment of the cell.

An enzyme, shown here colored yellow, has a cleft that only accepts molecules with shapes that fit the cleft (a). Once a molecule is in the cleft (b), a chemical reaction takes place. In some enzymes, such as the one shown here, the reaction helps break up the molecule (c and d); in others, the enzyme helps to join molecules together. Each type of enzyme has its own unique shape and is able to act only upon a specific range of molecules.

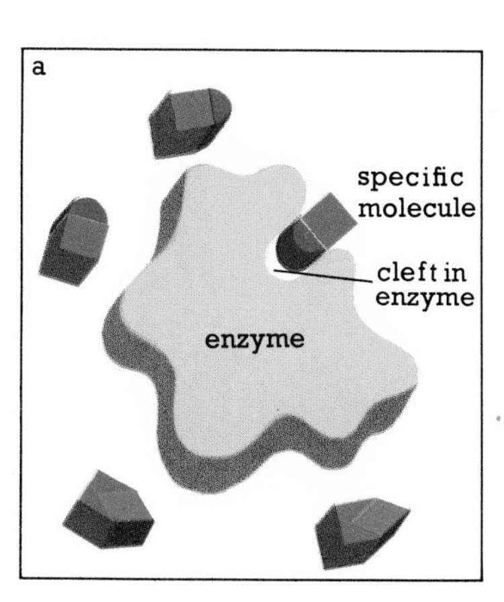

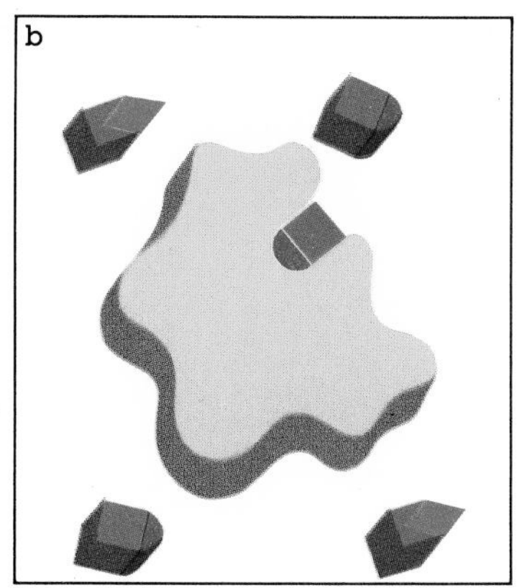

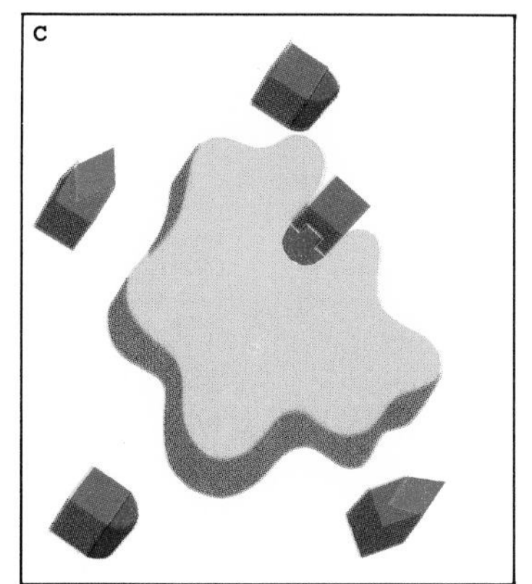

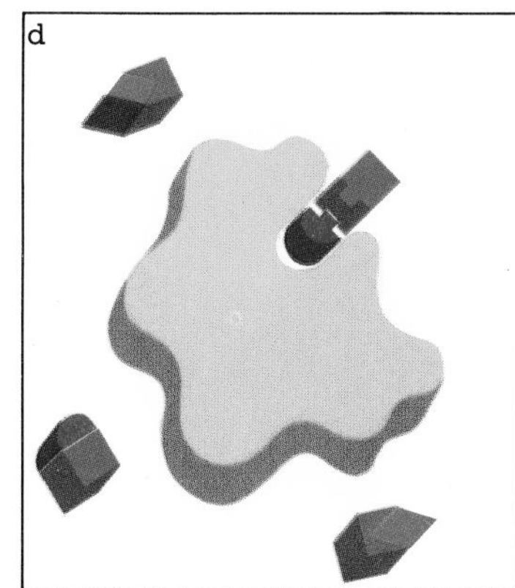

In Search of Genes

The first crucial step toward identifying the chemical nature of the gene was taken in a rather unlikely way. In 1928, Frederick Griffith, a British scientist, was studying the bacteria responsible for pneumonia, in the hope of finding some means of preventing or curing what was a common and often fatal disease. He had several varieties of bacteria and was able to grow them by injecting them into mice, which eventually died as a result. Pneumonia bacterial cells are usually surrounded by a coat, composed of a shiny material, giving them a smooth appearance, but Griffith began to find occasional cells that lacked this coat and looked relatively rough. These rough varieties grew and reproduced to give further rough bacteria, which Griffith suggested were mutants of the more common smooth bacteria. Mice injected with the smooth-coated bacteria died, while those injected with rough-coated bacteria did not. It appeared that the smooth bacteria were protected by their shiny coats against attack by the mouse's immune system, its defense against foreign bacteria and viruses.

Griffith found that if he first killed the normally lethal smooth bacteria by heating they, too, became harmless. He then discovered that if he injected a mouse with a mixture of dead, smooth bacteria and live, nonlethal rough bacteria, the mouse developed pneumonia and died. When he examined the dead mouse, he found its blood contained living smooth bacteria, and those smooth bacteria could grow and reproduce to give more smooth bacteria. It seemed that during their stay together in the mouse the live, rough bacteria had been transformed by the dead, smooth ones. Through its cell wall, the living bacteria had absorbed material that caused a structural change, and that could be passed on to descendants – in other words, a gene.

The next step was to find out which part of the bacteria was the gene for a shiny coat. The American scientist Oswald T. Avery and his colleagues at the Rockefeller Institute in New York, continued the work Griffith had begun. Avery broke open cells of smooth pneumonia bacteria, extracted their chemical components and tested each component separately for its ability to transform other, rough, pneumonia bacteria into smooth ones. It took the researchers several years to purify sufficiently the chemicals from smooth bacteria and to be fairly sure that they had found the one of which the gene was made. That chemical is called deoxyribonucleic acid, commonly abbreviated to DNA.

The next step was to find out whether all genes were composed of DNA or whether it was just true of genes for smooth coats in pneumonia bacteria. Work by Avery's group soon showed that in pneumonia bacteria, genes for characteristics other than smooth coats are also composed of DNA. It was already known that DNA, or in the case of some viruses a

These photographs show small mounds, or colonies, each of which contains millions of pneumonia bacteria, growing on the surface of jelly containing nutrients. In the photo at left are colonies of harmless rough bacteria, while those at right contain lethal smooth ones.

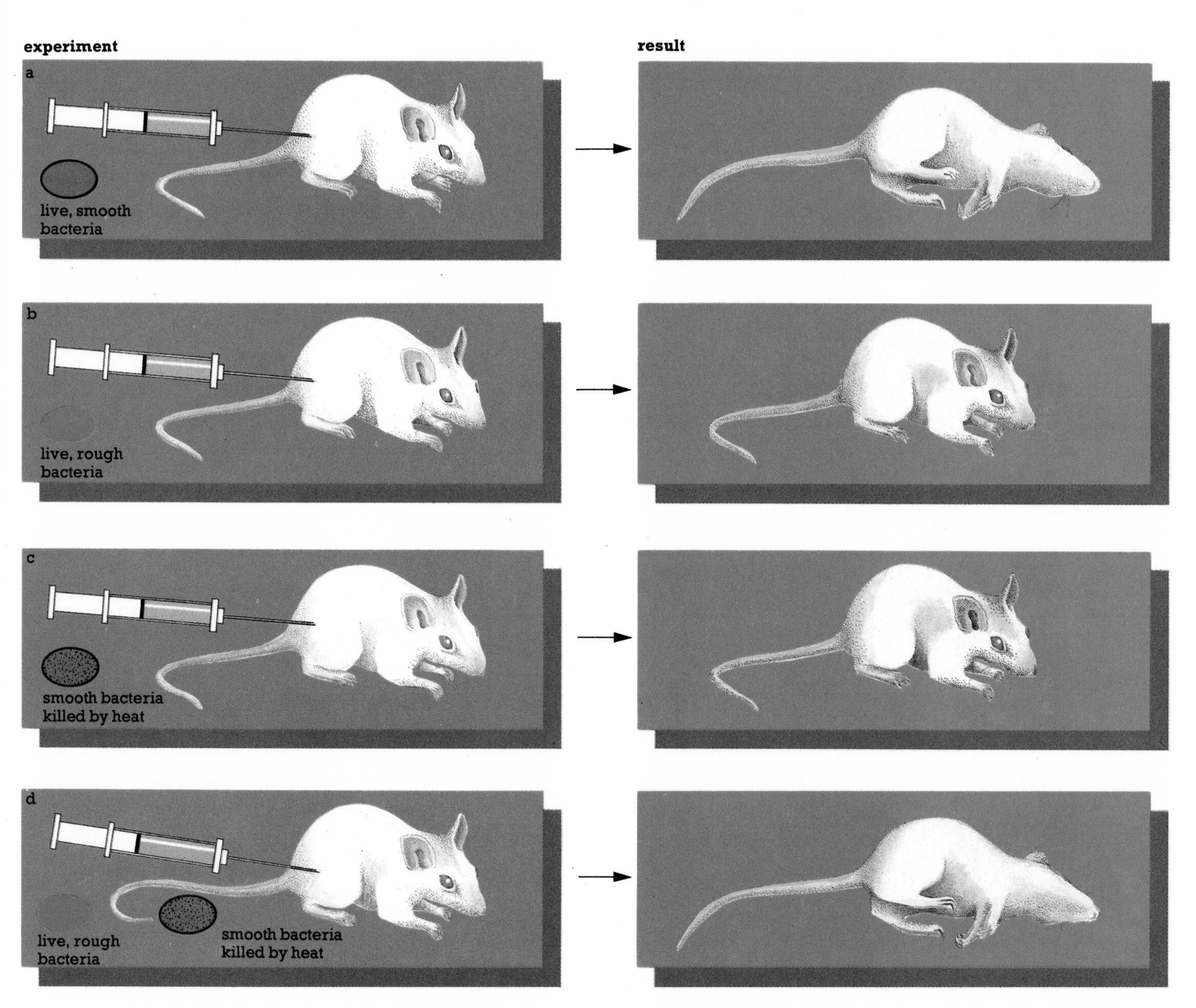

closely related compound called RNA (ribonucleic acid), occurs in the cells of all living things, and is passed on from cell to cell when cells divide. Eggs and sperm also contain DNA. So DNA is passed on from generation to generation of cell and thus from generation to generation of organism by sexual reproduction. This knowledge eventually led researchers to conclude that DNA (or RNA) is the material of which genes are made, not just in pneumonia bacteria, but throughout the entire living world.

Mice are commonly used in experiments because they breed quickly and because their internal organs and the way in which these function are basically similar to humans. This experiment shows the effects of different kinds of pneumonia bacteria on mice. An injection of live, smooth pneumonia bacteria is lethal (a), while live, rough bacteria are harmless (b). Smooth bacteria killed by heat are also harmless (c), but become lethal when mixed with the live, rough variety (d). This experiment showed that dead, smooth bacteria contain a substance, later identified as DNA, that transforms live, rough ones.

The Structure of Genes

DNA, the versatile material that makes up the gene, owes its ability to issue instructions for producing enzymes and other proteins to its remarkable structure. The complex chemical makeup of DNA gives it a wideranging effect combining the flexibility of an encoding machine and the predictability of a computer. DNA is one of the largest known molecules: the weight of one molecule of human DNA, for example, is about 100,000 times more than that of one molecule of sugar. Of course, in everyday terms, a DNA molecule is still very light: one quintillion molecules weigh only about three ounces. The DNA molecule is also very long, and if stretched out the DNA in a human diploid cell would be about six feet long, several thousand times the length of the cell that contains it! Inside the cell, the molecules of DNA are crammed in, tightly folded up.

The bacterial cell in the center has been deliberately burst. The cell thus releases its single molecule of DNA, whose long strand, coiled so that it resembles a vast maze, fills the rest of the picture.

Unfolded, DNA can be seen to consist of two long thin strands which are wound around each other to form a compact spiral. This spiral is called a helix and because there are two stands of DNA, the DNA molecule is referred to as a double helix. The strands of the double helix are made up of a very large number of individual units, or chemical building blocks, called bases. These are linked together by chemical bonds, as beads are strung together on a necklace. There are four types of bases in DNA, differing in size and precise shape: the chemicals guanine and adenine, abbreviated to G and A, are the two larger bases, and the two smaller ones are cytosine and thymine, abbreviated to C and T. Each strand of DNA contains up to twelve billion individual bases, and since the four types A, G, C, T, occur in various sequences many different arrangements of bases can occur.

The two strands of the DNA helix are held together by a series of weak chemical bonds between individual bases in adjacent strands. The combination of the regular turns of the helix and the shapes of the bases means that the large base G only pairs up with a C opposite in the other strand (and a C with a G). Likewise the other large base A can only pair with the smaller T (and T with A). This means that no matter how complex the sequence of bases in either of the two strands of a DNA molecule, this sequence will be mirrored in the other strand. For example, if a portion of one strand of a DNA molecule comprises AGCCTATGG, the other will have the sequence TCGGATACC. Thus, in scientist's language, the bases A and T, and G and C, are complementary to each other, and the double helix DNA is composed of two complementary strands.

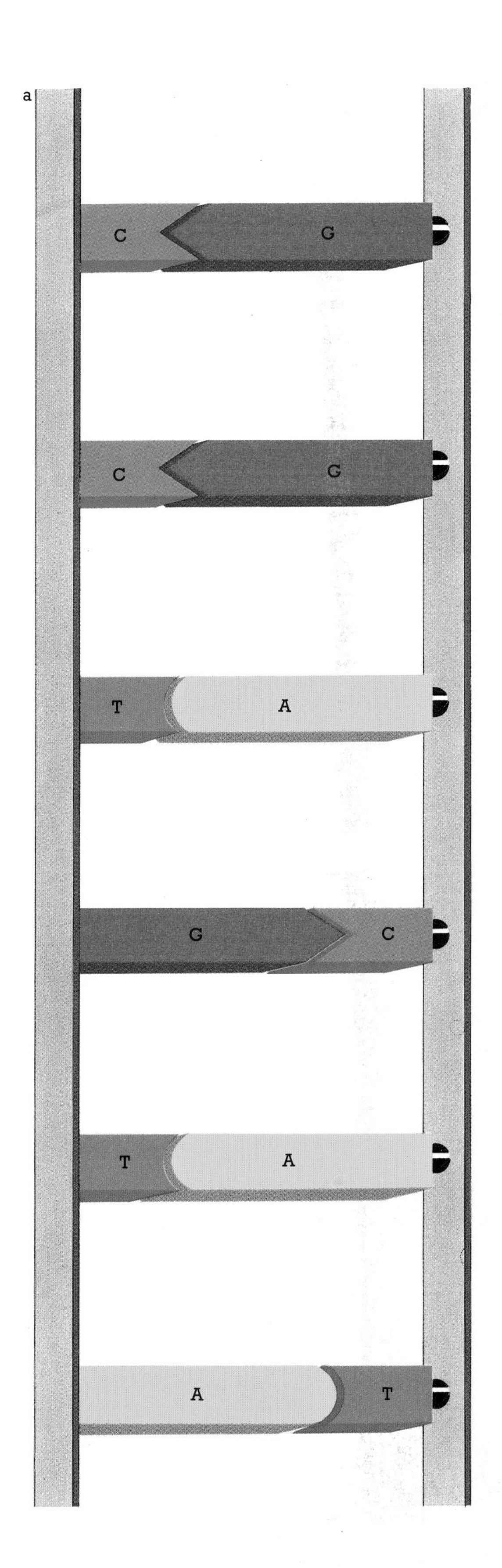

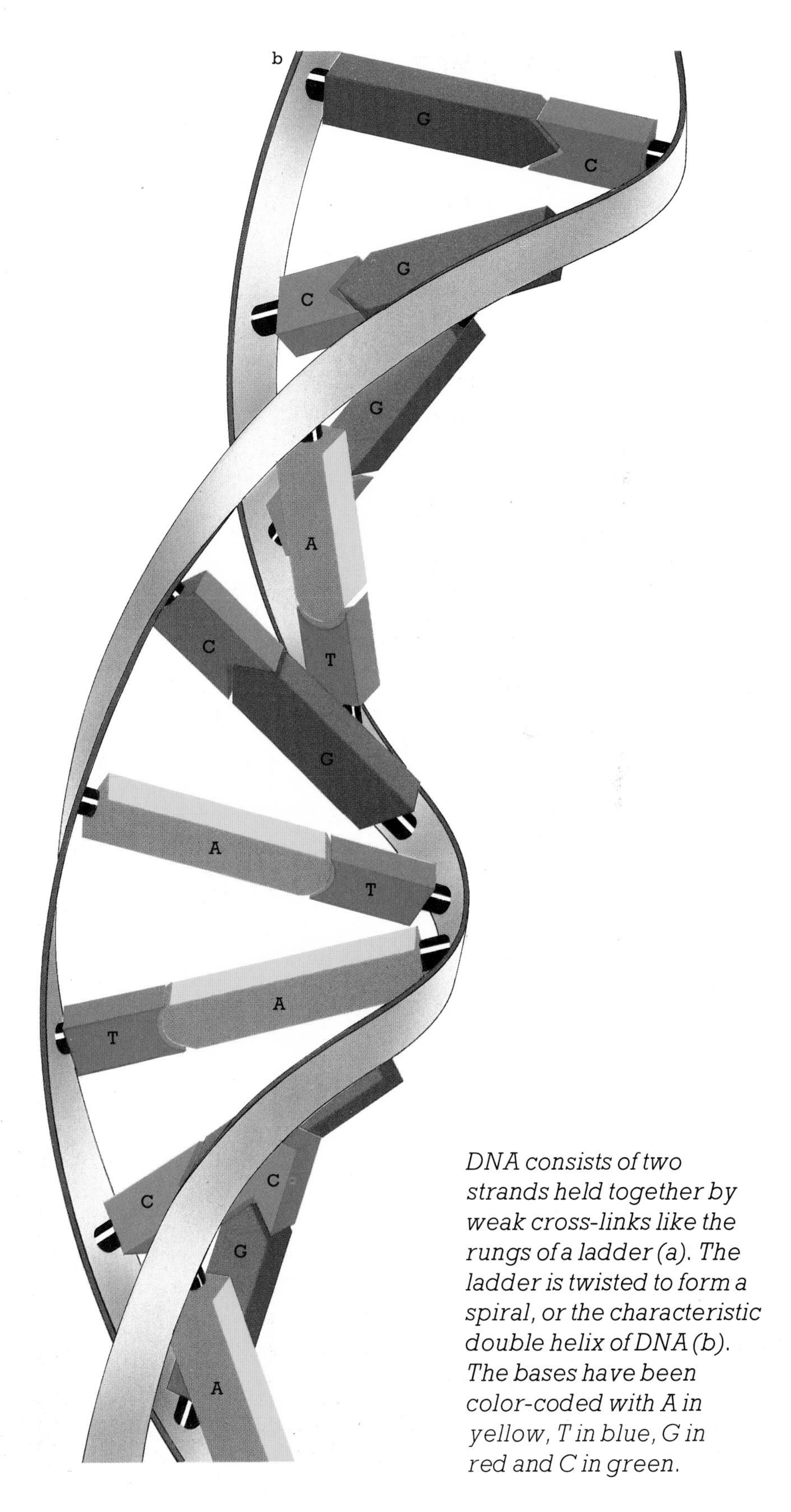

DNA consists of two strands held together by weak cross-links like the rungs of a ladder (a). The ladder is twisted to form a spiral, or the characteristic double helix of DNA (b). The bases have been color-coded with A in yellow, T in blue, G in red and C in green.

DNA and the Genetic Code

The sequence of bases in the strands of DNA is the key to the double helix's secret as an encoder. Each DNA base is like a letter of the alphabet, and so a sequence of bases can be regarded as forming a message. These messages are the genes. The average gene consists of a sequence of hundreds or thousands of individual bases. So, although only four bases exist in DNA, the number of possible sequences, and therefore genes, is enormous.

Each gene instructs the cell that contains it to make a specific protein (such as an enzyme) by dictating a particular sequence of amino acids, the building blocks from which proteins are made. However, in proteins there are twenty different types of building block, not just four. So the problem is this: how does a gene, whose information is contained in the sequence of the four types of bases, determine the ordering of twenty different types of amino acid in a protein?

The problem can be solved by noting how in Morse code all twenty-six letters of the alphabet can be encoded with just two symbols – the dot and the dash. The trick is to use different combinations of dots or dashes to code for single letters of the alphabet. The cell employs a similar method, using differing combinations of bases. Each group, called a codon, is made up of three of the four bases available. Altogether there are sixty-four codons, many more than the twenty needed to specify the twenty amino acids. Sixty-one of these codons are actually used to code for the twenty amino acids, which means that

The structure of DNA was first proposed in 1953 by the American biochemist, James Watson (left), and the British molecular biologist, Francis Crick (right), seen here with an early model of the DNA molecule. Crick has since made major contributions toward unraveling the genetic code, while Watson has contributed to our understanding of the structure of RNA.

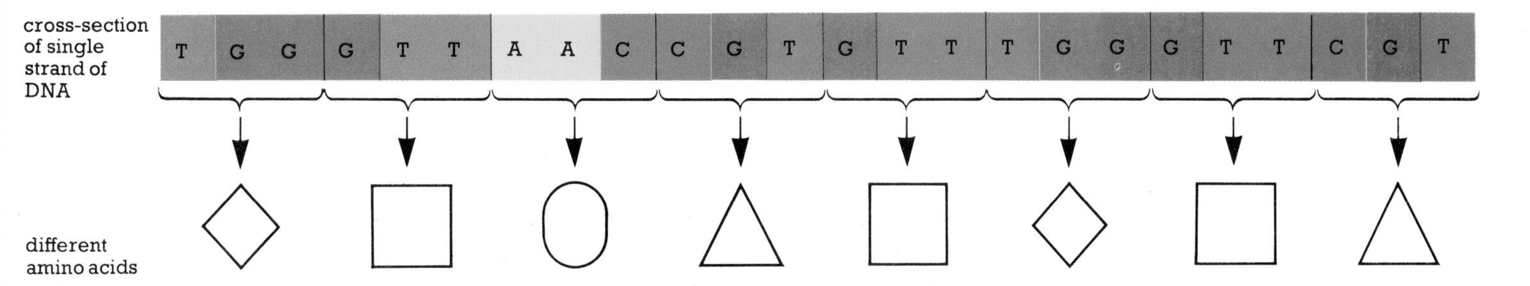

In the genetic code the sequence of bases is read from a fixed starting point, just as a sentence is read from the beginning. Here the bases (top row) are lettered as well as color-coded to represent thymine, guanine, adenine and cytosine. Bases are grouped in threes to form codons that specify different amino acids, shown as squares, triangles, diamonds and ovals in the bottom row.

A researcher uses a glass rod to collect long fibers of DNA that have separated out in a chemical solution. This is one step in the extraction and purification of DNA from living cells for experiments in the commercial applications of genetics.

some amino acids can be specified by more than one type of codon.

The cell reads one codon at a time, and for each codon read a specific amino acid is joined to a growing sequence of amino acids that will form a protein. However, just like a book with its gaps, periods, paragraphs and chapters, the DNA needs punctuation to convert a continuous sequence of hundreds of thousands of bases into a series of precise instructions. The remaining three of the sixty-four codons act as periods, telling the cell that a gene has finished producing its chain of amino acids. In addition, there are probably quite long sequences of bases between genes that do not code for proteins, sequences that may say "I'm not a gene," or "I'm not a gene but one is starting soon."

Building Proteins

Reading instructions on how to do something and actually doing it are two very different things. If we were building a car an instruction manual would be very useful, but we would also need materials, tools, probably several people, and time. DNA acts as the cell's instruction manual for building numerous types of protein and, like a car builder, the cell, too, needs tools and materials to carry out the instructions.

Because one person could not efficiently make every component of a car, people making the various parts of the car would use different sections of the instruction manual – the section on how to construct a steering wheel would be used by one worker, that on making headlights by another, and so on. In a somewhat similar way, the cell does not use all the instructions contained in DNA at once, but uses particular sections, that is particular genes, as needed, with different proteins representing the cell's "steering wheel," "headlights," and so on.

In all organisms other than bacteria and their close relatives, the DNA is stored wrapped up in the nucleus of the cell. When a cell needs a particular protein and so needs the instruction contained in a particular gene, just that instruction is dispatched to the ribosomes in the cytoplasm. Ribosomes are a vital part of the cell's machinery because they respond to an instruction by assembling the proteins called for.

Just as photocopy sections of the instruction manual would be handed out to the different workers, rather than tearing sections out of the book for distribution, so individual genes are not themselves dispatched to the cytoplasm. Instead, copies of the instructions contained in genes are sent out. These copies are in the form of molecules of RNA, which is similar to DNA but has just one strand of bases instead of two and the base uracil (U) instead of thymine (T). The RNA copies are formed by an enzyme that can be called messenger-maker. It puts together a sequence of bases that are complementary to just one strand of the two in the DNA of the gene in question.

When the RNA copy, called messenger RNA,

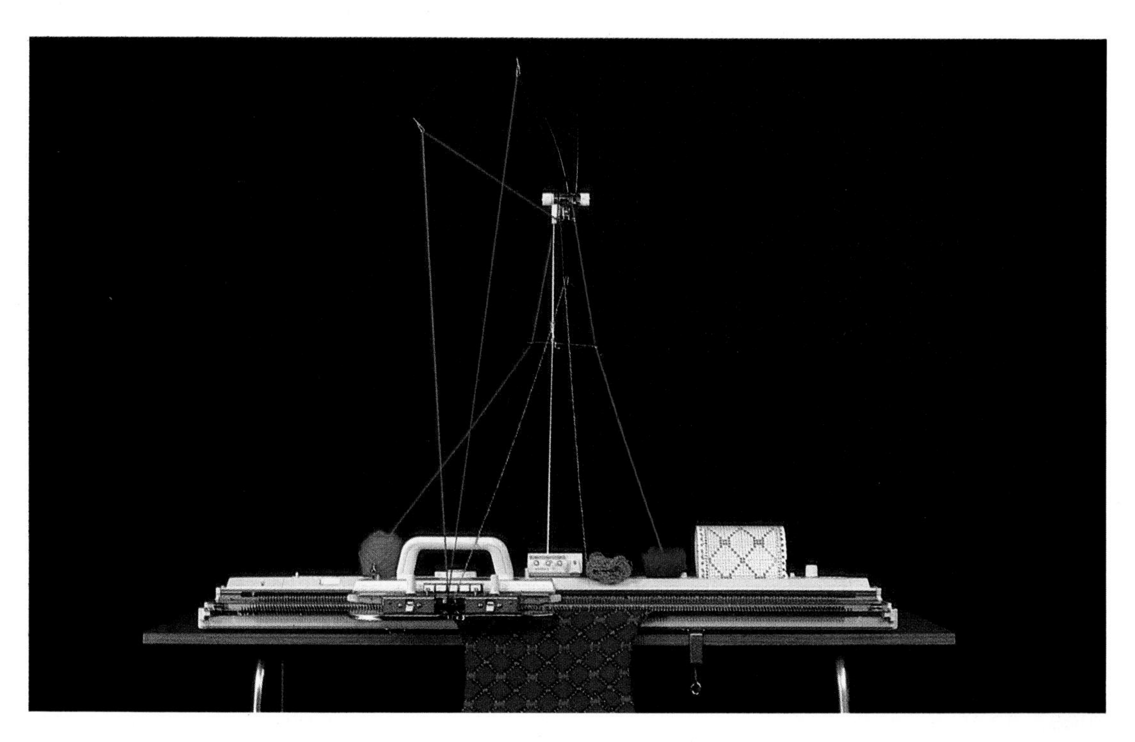

This knitting machine can be programmed to knit a garment to any required pattern. The machine is provided with a series of punched cards. The position of the holes in the card determines what color and type of stitch will be produced at each step. Proteins are assembled in much the same way. They consist of chains of amino acids that are themselves produced by the decoding of messenger RNAs. Sequences of three bases (codons) determine which type of amino acid will be produced.

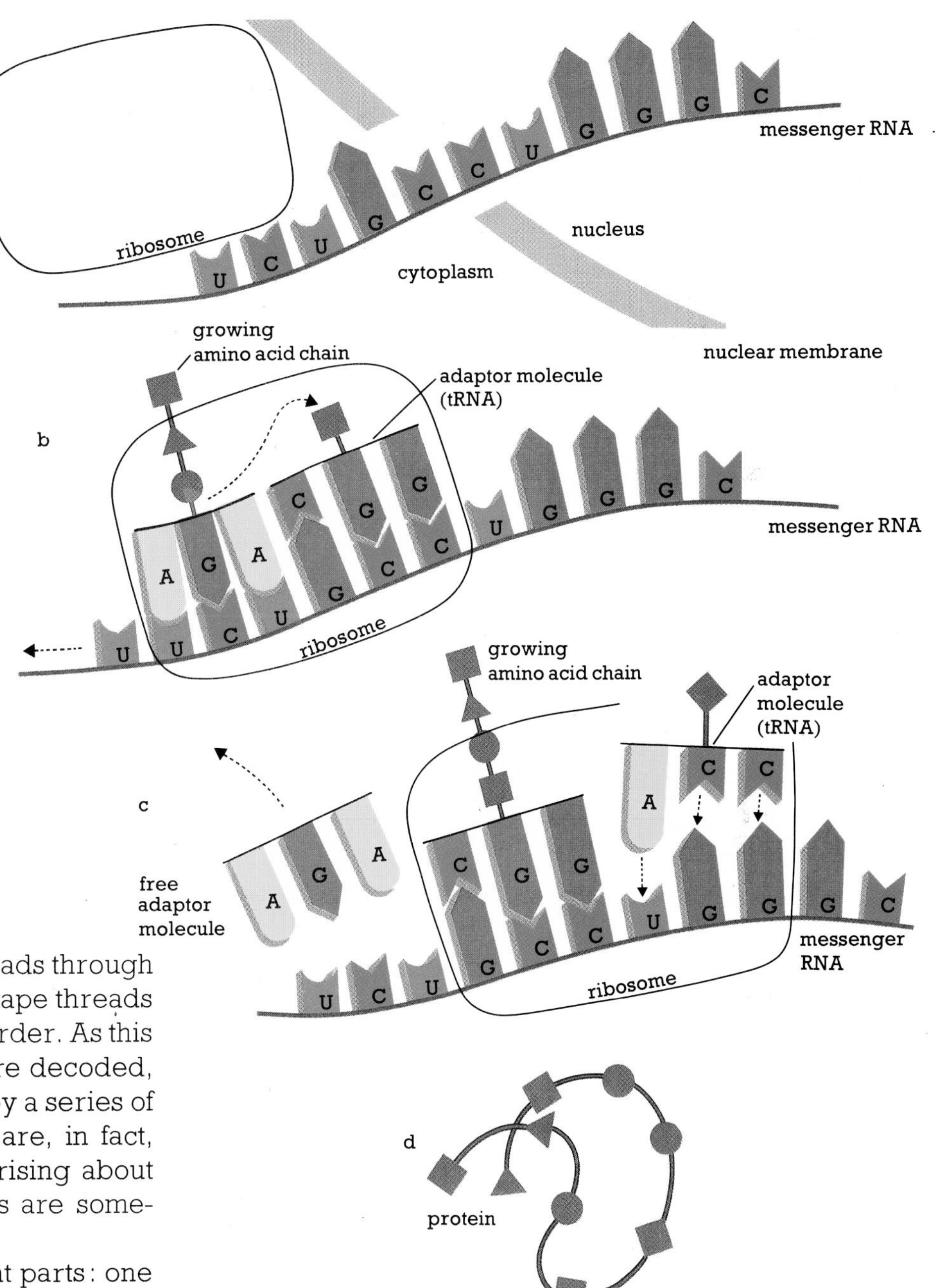

In this schematic diagram, a messenger RNA, or mRNA, which is a single strand of bases, has been copied from a gene and passes from the nucleus to the cytoplasm toward a ribosome (a). As the mRNA is threaded through the ribosome, it is decoded by adaptor molecules, called transfer RNA, or tRNA, one codon (three bases) at a time to give an amino acid chain. At each decoding step the growing amino acid chain is transferred onto the adaptor molecule carrying the amino acid for the most recent step (b). The free adaptor is ejected and the next one approaches (c). When the end of the message is reached, the completed amino acid chain separates from the ribosome and folds up into a protein (d).

arrives in the cytoplasm of the cell, it threads through a ribosome similar to the way in which a tape threads through the playback head of a tape-recorder. As this happens it is read, and its instructions are decoded, three bases, that is one codon, at a time by a series of adaptor molecules. Adaptor molecules are, in fact, themselves small RNA molecules comprising about 75 to 90 bases. These special molecules are sometimes called transfer RNA molecules.

An adaptor molecule has two important parts: one carries a particular amino acid, while the other fits a particular codon of RNA. Each specific adaptor carries just one of the twenty types of amino acid. So the adaptors help match codons with their correct amino acids.

Whenever two adaptors lie side by side in the reading head of the ribosome, their amino acids become chemically linked to each other. The adaptor attached to the earlier codon is then ejected, and the messenger RNA winds on one codon through the ribosome. The new codon in the ribosome reading head now attracts the next appropriate adaptor with its amino acid in tow, which is joined to a growing chain of linked amino acids. Eventually a period codon signaling "end of message" slots into the reading head. The completed chain of amino acids then separates from the ribosome and folds up to give the protein its characteristic shape.

Switching Genes On and Off

Making proteins is an important feature of growth and renewal. Every cell contains a large variety of different proteins, some of which are structural while others are enzymes. Many of the proteins are needed in the cell's day-to-day chemistry – what is sometimes called the "housekeeping" of the cell and the genes coding for these proteins produce messenger RNAs almost continuously. However, other proteins are not involved in general housekeeping, and are needed only occasionally. Although the genes coding for these proteins are always present in the cell, they produce messenger RNAs only when the proteins are needed. In other words, such genes can be switched on and off.

Even though virtually all cells are capable of switching some genes on and off, much that is known about the process comes from the study of single-celled organisms, bacteria (the singular of which is bacterium). One type of bacterium, in particular, has been studied by geneticists. This is *Escherichia coli* (*E. coli* for short), which normally lives inside the intestine, but which can also be grown easily and cheaply in a laboratory.

Placed in a solution of salts and glucose, *E. Coli* can grow and reproduce by simple cell division once every thirty minutes or so. This bacterium is an amazingly versatile chemist and employs a wide range of enzymes to use the carbon contained in glucose to help make many complicated molecules.

E. coli will also grow if glucose is replaced by lactose, a sugar found in milk. To use lactose, *E. coli* first converts part of it to glucose. To do this it needs a specific enzyme, which is only present in minute quantities when *E. coli* is growing on glucose. Within

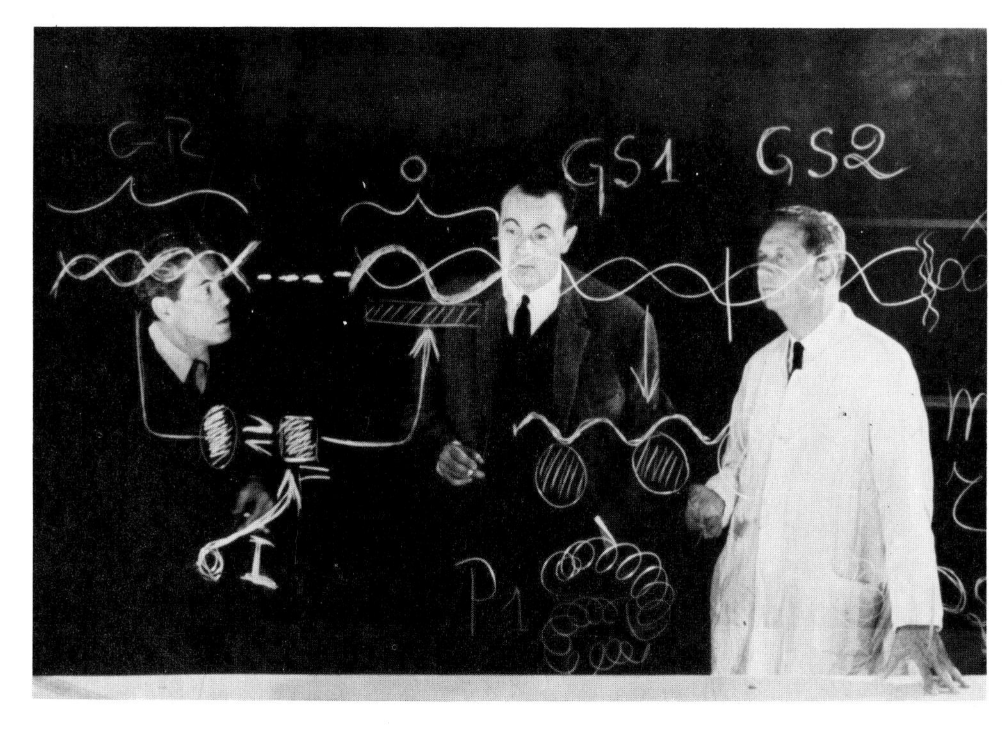

Standing behind their transparent "blackboard" are Jacquès Monod (left), François Jacob (center) and André Lwoff (right), who shared the Nobel Prize for Physiology or Medicine in 1965. Monod and Jacob were largely responsible for our understanding of the production of the lactose-converting enzyme in E.coli. *Lwoff increased our knowledge of the way in which certain viruses can live inside* E.coli.

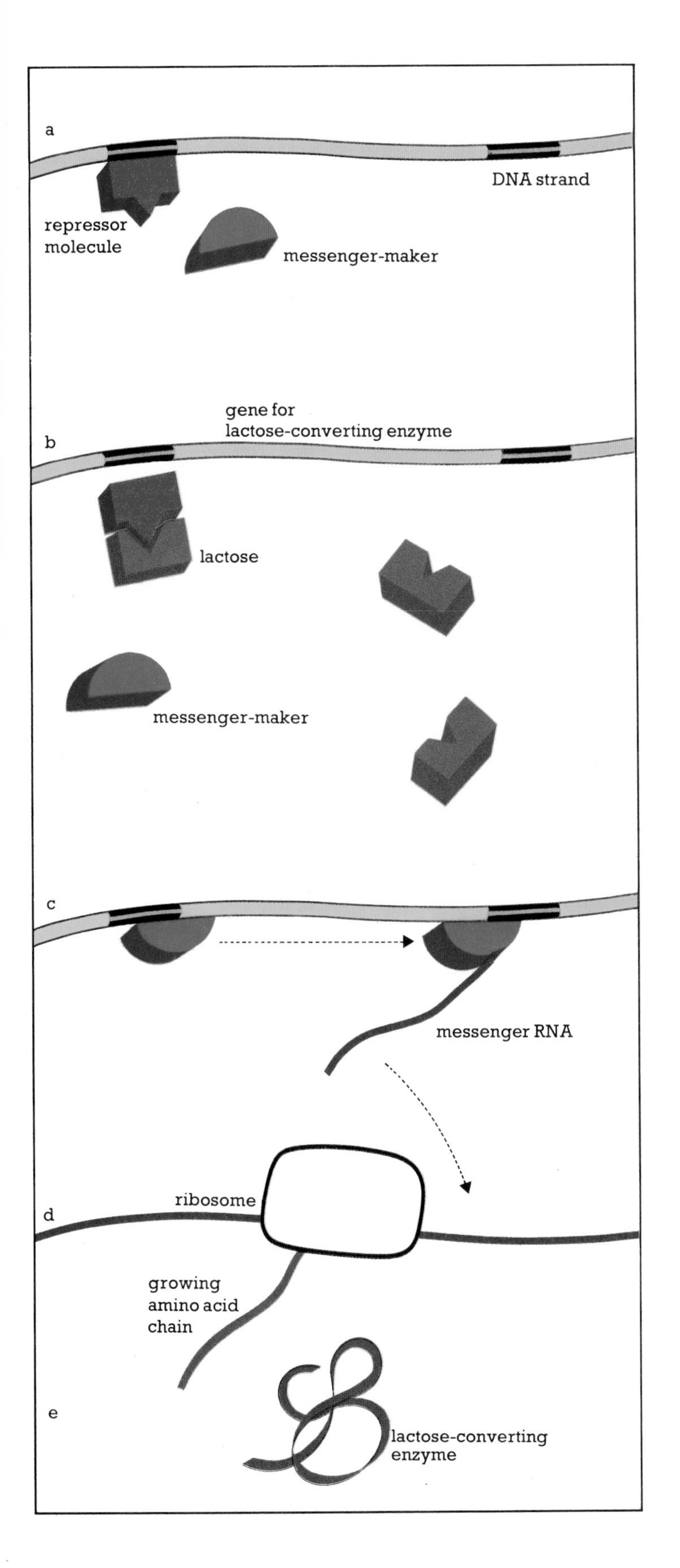

three or four minutes of being presented with lactose in place of glucose, *E. coli* starts to make large quantities of this lactose-converting enzyme. The organism does this by switching on a gene that contains instructions for making the enzyme. The gene becomes switched on when the so-called messenger-maker enzyme, enables the DNA to produce the messenger RNA.

When *E. coli* is growing on glucose, and lactose is absent, messenger-maker is prevented from reaching the gene for the lactose-converting enzyme by a special molecule, called a repressor. The repressor attaches itself to the DNA near the starting point of the gene. However, when lactose molecules are present, one of these molecules fits onto the repressor, causing it to fall off the DNA. Then messenger-maker can begin making messenger RNA.

This "on-off" switching system is just one example of many similar means for controlling the rate at which proteins are made. Each system uses a specific repressor, with its own particular position on the DNA, and this can be removed only by one substance specific to the repressor. The cell is, therefore, able to produce a particular enzyme for just as long as it is needed.

In this schematic summary of a genetic control system, a small portion of the DNA of E. coli *is shown in yellow, with the region just before the gene for lactose-converting enzyme highlighted in black. When* E. coli *is growing on glucose, and not on lactose, a repressor molecule (red) prevents the messenger-maker (blue) from reaching the start of the gene for lactose-converting enzyme (a). When present, lactose (green) binds onto the repressor and causes it to fall off the DNA (b). This enables messenger-maker to reach the start of the gene and begin copying it (c). The copying produces messenger RNA, shown as a blue thread, which is then read in the ribosome (d) to produce lactose-converting enzyme (e).*

Making DNA

Much more common than sexual reproduction, in the world of living organisms, is a means of perpetuating the species that does not involve sex. Asexual reproduction is a constant process in the body as individual cells reproduce themselves.

Like any complex piece of machinery cells occasionally need replacing. This replacement occurs when a healthy cell divides to give two daughter cells. Normally, before it divides a cell grows by taking in foodstuffs. The cell converts these foodstuffs into substances and structures similar to those it already contains. In this way, the cell can double in size before division. If the cell did not take in food and division still took place, the cell would get smaller with each division. The two daughter cells resulting from the first division would divide to give four cells, then eight, and so on, until they disappeared!

When it divides by asexual reproduction the cell has to ensure that both daughter cells receive a complete set of genes, like those of the parent cell, so as to make sure the same set of genes is passed through every cell generation. To do this, two things happen. Before it divides the parent cell makes a copy of all its genes; that is, the cell reproduces its own DNA so that it now has two complete sets. Then, when division occurs, the cell distributes its two sets of genes equally so that each daughter cell has a complete set just like those of the parent.

The reproduction of genes takes place in two stages. DNA consists of two strands. The bases in one strand are complementary to those of the opposite strand. The chemical bonds that join the bases together in each strand of the DNA double helix are strong, but the bonds pairing A-T, and G-C in adjacent strands are weak. These weak bonds can break so that the double helix of DNA gradually comes apart.

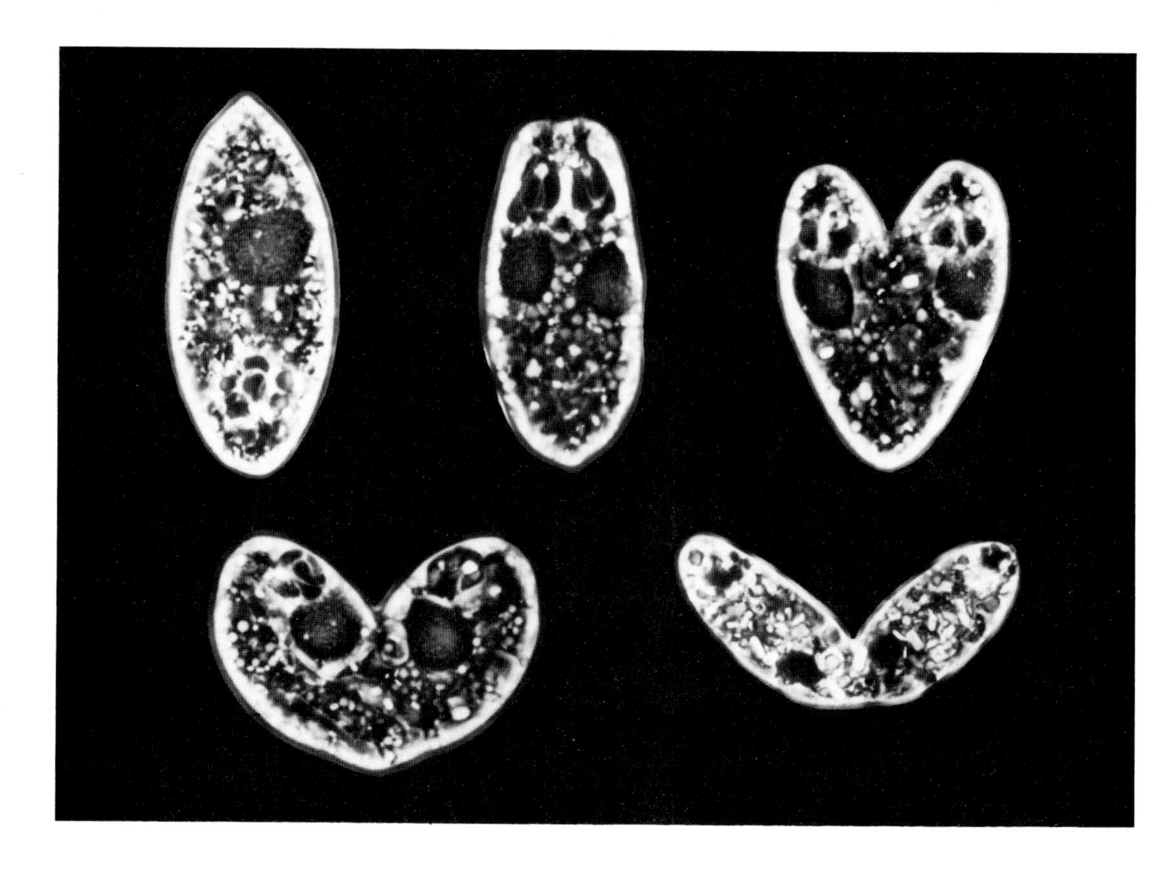

Euglena, *a single-celled, plant-like organism, undergoes progressive stages of cell division.*

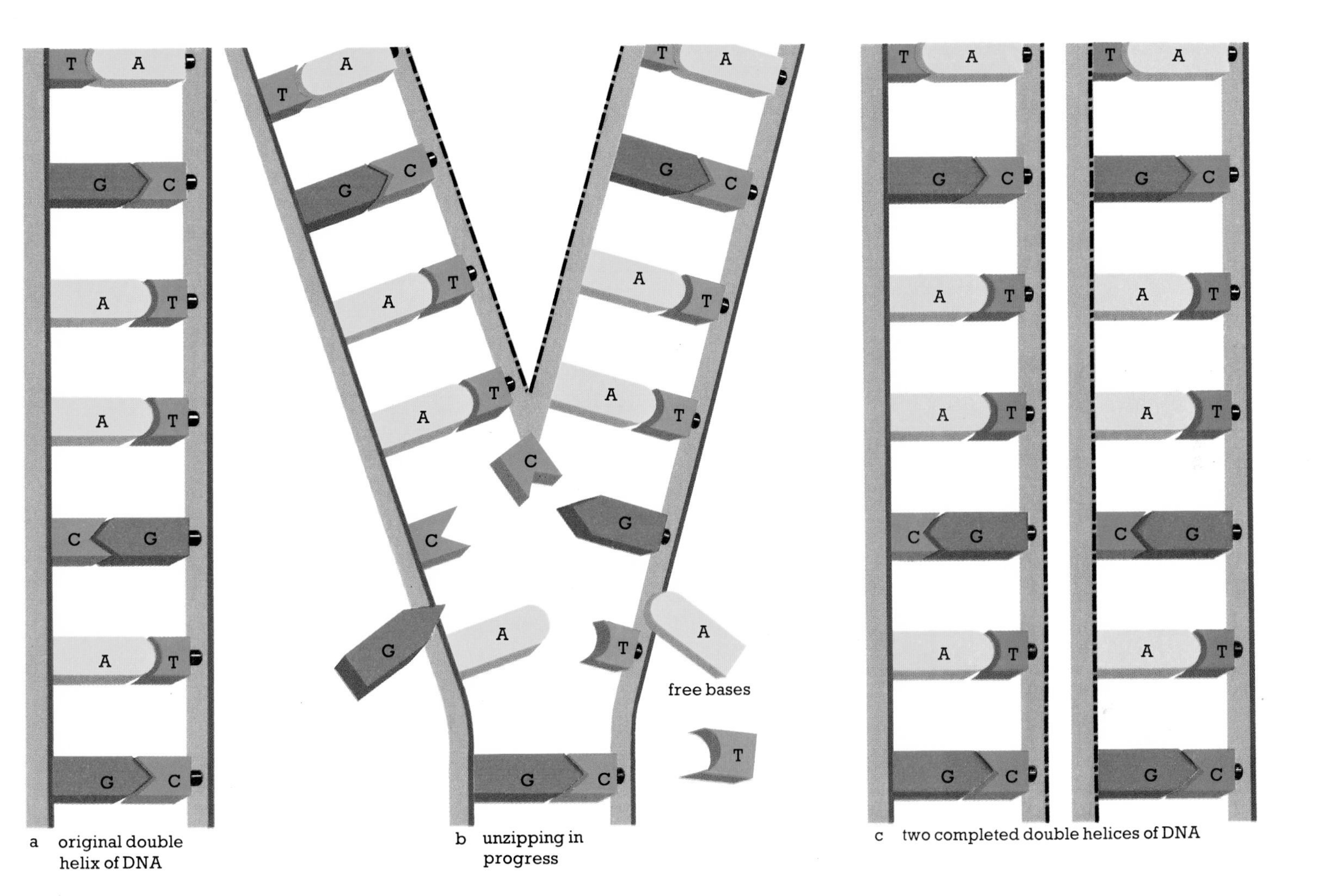

As it unzips, each of the two strands acts as a kind of mold for forming a new complementary strand.

There are many unlinked bases lying in the cell like loose beads. These form the new strands of DNA. The sequence in which they are joined is determined by the sequence of the two unzipping strands of an existing double helix. Each unzipping strand attaches bases parallel to itself in an order appropriate to its own sequence: A with T, and G with C. The result is two new double helices (the plural form of helix), in each of which one strand is from the original double helix and one strand is formed of newly joined bases. And, of course, each of the two double helices has exactly the same sequence of paired bases, as in the original double helix.

These simplified diagrams show the steps involved in making DNA. The double helix (a) reproduces when it unzips and each strand gradually builds up a strand complementary to itself (b). This happens when bases, from the stock always present in the cell, attach themselves to the appropriate bases on the unzipping strands. These incoming bases are joined together to form new strands. This eventually results in two complete double helices of DNA (c). Each helix comprises one strand from the original double helix, shown at (a), and one completely new strand, in which the bases are shown linked by the broken line.

Mitosis

Cells contain some very sophisticated machinery for distributing DNA equally between daughter cells. In animal or plant cells DNA is contained in the cell nucleus, where it is wound around spherical clusters of proteins called histones. The resulting structures form chains. These are in turn coiled up to form chromosomes, each of which contains a single long double helix of DNA.

Every species of animal or plant has a fixed and characteristic number of chromosomes per cell. Most of the time chromosomes cannot be seen even under a powerful microscope. Moreover, the cell nucleus has a speckled, rather dull appearance. Although it may appear uninteresting, the nucleus is actually a hive of activity, because DNA is being reproduced. The new strands of DNA are also wrapped around histones to form chromosomes, so each chromosome now has an identical twin. Together they form a tightly knit pair. The stage is now set for cell division and distribution of the chromosomes containing DNA.

Just before the cell divides, the chromosome pairs shorten and thicken and become visible under the microscope as sausage-like structures. These line up along the center of the cell. Each pair of chromosomes begins to split up and its two members move toward opposite ends of the cell. This very precise and rhythmic movement, sometimes nicknamed the "dance of the chromosomes," is coordinated by a series of fine threads of protein. These threads act as

The pictures (below) illustrate phases in mitosis in cells in the root of an onion. In (a) the nucleus of the box-like cell in the center is in its apparently inactive phase, where no chromosomes are visible. In (b) chromosomes become visible, sort themselves out and line up about the center of the cell (c). They then separate towards opposite ends of the cell (d and e). By (f) they have reached opposite ends of the cell, have become less distinct, and two new nuclei are forming. A new cell membrane is being laid down between the two nuclei (g). By (h) two new complete daughter cells have formed.

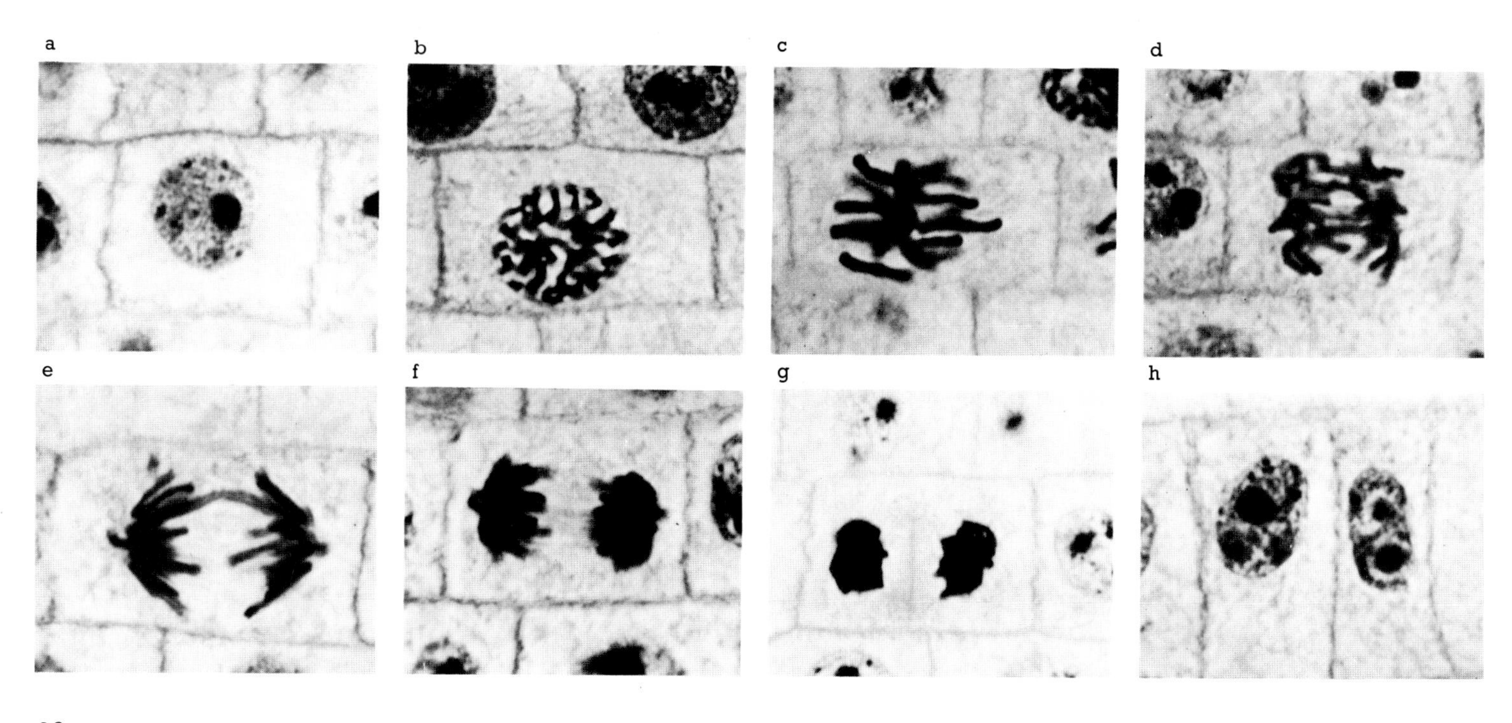

Asexual reproduction has allowed algae to colonize rapidly on the surface of this canal.

a kind of cat's cradle to help pull the chromosome pairs apart. When they reach opposite ends of the cell, each set of chromosomes becomes surrounded by a double membrane to form two separate nuclei. Finally, a membrane forms across the center of the cell to create two new cells, each with a similar set of chromosomes. This process of DNA reproduction followed by cell division is called mitosis.

Bacteria, which consist only of a single cell, also multiply by asexual cell division. Their DNA is not wrapped around histones, and so they do not have any structures quite like the visible chromosomes of higher organisms. Nevertheless they also accurately pass on a complete set of genes to each daughter cell by a primitive form of mitosis.

In most higher multicellular organisms the role of asexual cell reproduction is limited to the growth and renewal of tissues. However, some animals and plants can produce an entirely new organism by asexual reproduction. For example, strawberry plants can, by cell division, produce a horizontal stem or runner, which then develops into a complete plant. Similarly, some parasitic worms can reproduce asexually simply by breaking into eight or more pieces. Each piece may then mature to give a whole worm!

Sex Cells and Reproduction

In organisms that reproduce sexually the number of chromosomes in a cell represents two sets. One set is a copy of those inherited from the organism's mother, the other is a copy of those from the father. Cells with a double set of chromosomes are said to be diploid. When cells divide, the daughter cells are also diploid, as the diploid set of chromosomes in the parent cells is doubled prior to division.

Sex cells – the eggs and sperm, or pollen – are a special case. These special cells converted from ordinary cells have half the number of chromosomes of other cells and are described as haploid. They have only one set of chromosomes each, so that when fertilization occurs the set from the male unites with the set from the female to form a single cell. This process creates the basic diploid cell again.

In converting from ordinary cell to sex cell – which usually takes place in a sex organ such as an animal's ovary or a plant's stamen – the number of chromosomes is halved in a complex process called meiosis. A quick and simple way to accomplish this would be simply to divide the total number of chromosomes in two. But that would not necessarily ensure that the sex cell contained one chromosome of each pair, adding up to a complete set. The cell puts its chromosomes through an intricate sequence of movements, which

Although they are sisters, the girls above do not look alike. It is not unusual for children of the same parents to be dissimilar in many obvious respects because genes play an important part in determining characteristics. Unless parents have identical twins, developing from the same fertilized egg, it is extremely unlikely that their children will have the same genetic makeup.

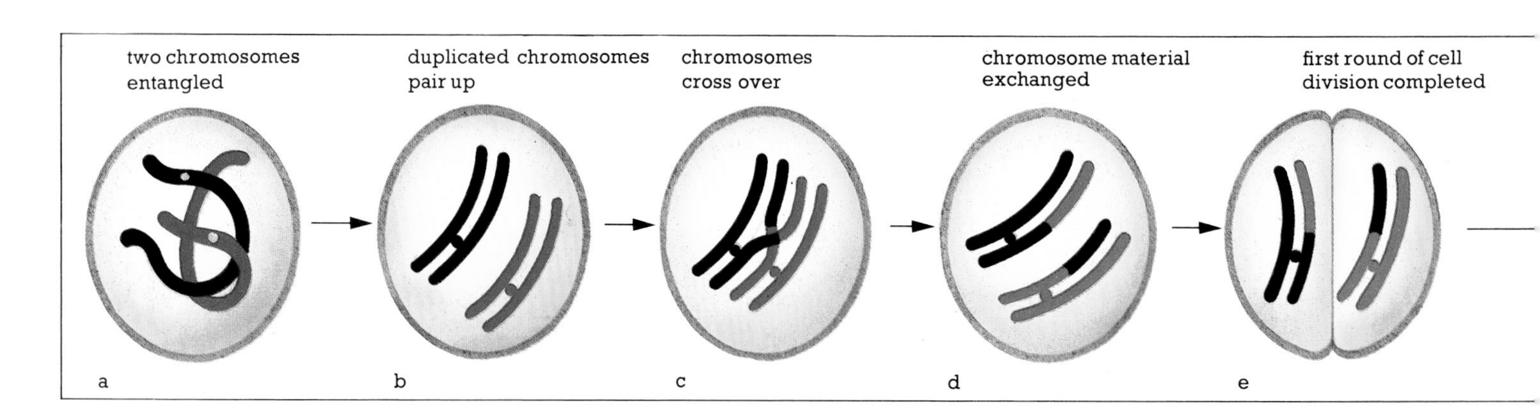

distributes them properly.

As meiosis begins, the chromosomes from the two parents, which are usually distributed at random, reproduce themselves and line up with each other to form matching pairs. Each pair actually contains four chromosomes – two copies of the mother's chromosomes and two of the father's.

Groups of genes initially derived from the parents are brought together in new combinations by a process of crossing over. Individual chromosomes can break virtually anywhere along their length and then rejoin. When a chromosome breaks, one of its counterparts from the other parent may also break at precisely the same spot along its length. The broken ends may then rejoin in a crossover so that each joins up with an end from its counterpart. Where each breaks and rejoins is a matter of chance, and this can occur in any of the groups of four similar chromosomes. The number of possible new combinations of genes is, therefore, absolutely enormous.

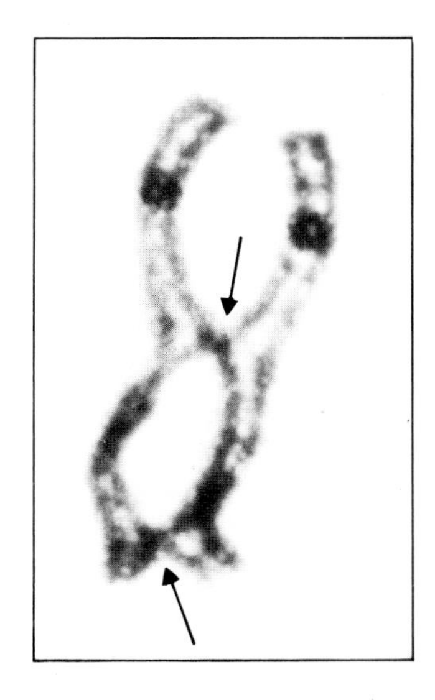

This photograph shows the crossover of the chromosomes in a salamander. In fact, two simultaneous crossovers are occurring here (indicated by arrows), further increasing the number of possible gene combinations.

The diagram below shows the behavior of one pair of chromosomes during meiosis. The red chromosome comes from the female parent and the black from the male parent. Chromosomes normally form a tangled mass within the cell (a). Before cell division the chromosomes untangle and reproduce themselves, then the red chromosome lines up with the black (b). A crossing over takes place (c), which results in an exchange of genetic material (d). The first round of cell division is completed (e), resulting in two diploid cells (f). Each duplicated chromosome splits (g) and the second round of cell division (h) results in four sex cells (i), each with half the number of chromosomes of the original diploid cell.

An organism generally produces a large number of sex cells during its lifetime. The chance way in which the chromosome pairs are shared during meiosis guarantees that any two egg, sperm or pollen cells are highly unlikely to have the same combination of chromosomes (that is, the same mixture of the mother's and father's chromosomes). For example, a human diploid cell has forty-six chromosomes, that is twenty-three pairs. The odds that two sex cells might receive the same mix of chromosomes are one chance in 8,388,608 even without considering the additional variations produced by crossing over. And since chromosomes contain the genes, the chance of two sex cells having precisely the same combination of genes is likewise very slim.

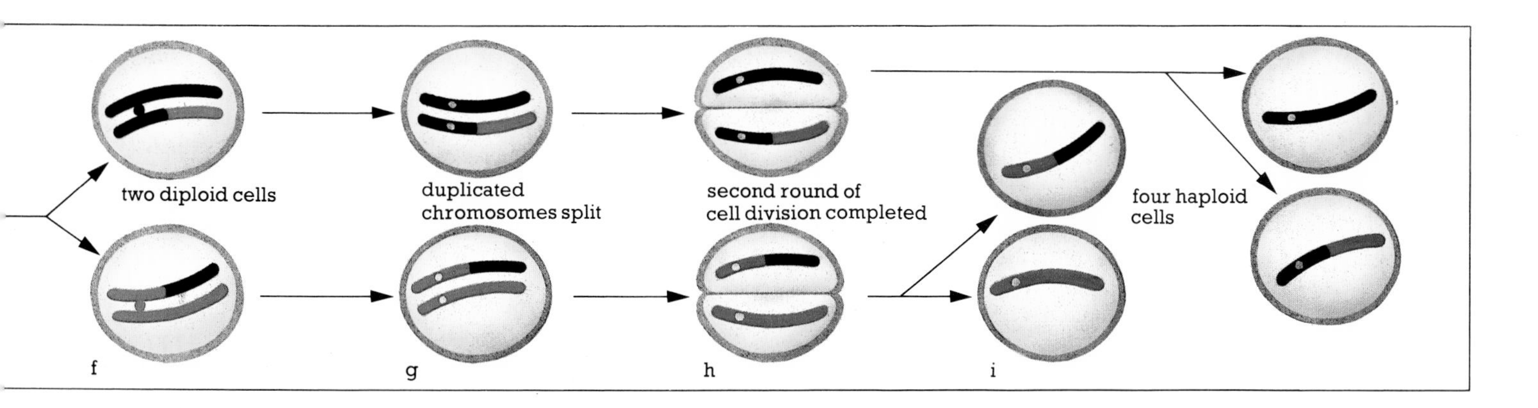

Sex Chromosomes

The chromosomes forming each of the pairs in a cell are very similar in appearance, with one important exception. If you carefully examine the appearance of human cells, other than those which are eggs or sperm, for example, you can see that the forty-six chromosomes of the cell fall into two categories: one has twenty-three pairs and the other twenty-two pairs with two odd ones. Cells with twenty-three pairs are a woman's while the twenty-two pairs plus the two odd ones belong to a man. The two nonidentical chromosomes from the male cells are called sex chromosomes, and because of their shapes one is called X and the other Y. A woman also has sex chromosomes, but these form a pair – two Xs.

Consider what happens at meiosis, when cells divide. The cells in a woman's ovary give rise by meiosis to egg cells, each of which contains twenty-three chromosomes. One of these chromosomes is an X. Likewise, in a man meiosis gives rise to sperm cells, each containing twenty-three chromosomes, of which one is either an X or Y. An X-containing egg may be fertilized either by an X-containing sperm, or a Y-containing one. If it is an X-containing sperm, the resulting egg will develop into a female; if it is a Y-containing sperm, then a male will be born. XX is female, XY is male.

Occasionally humans are born with abnormal sex chromosomes. For example, people may have just one X chromosome and no Y; these so-called X0 individuals are female, but do not mature sexually. People with an extra X – XXY – are underdeveloped

Dramatic moments in the fertilization of a sea urchin egg. In the main picture (magnification × 21,000), sperm swarm round the egg. At top right (× 21,000), sperm move onto the egg's surface. At bottom right (× 10,000), the sperm enters the egg.

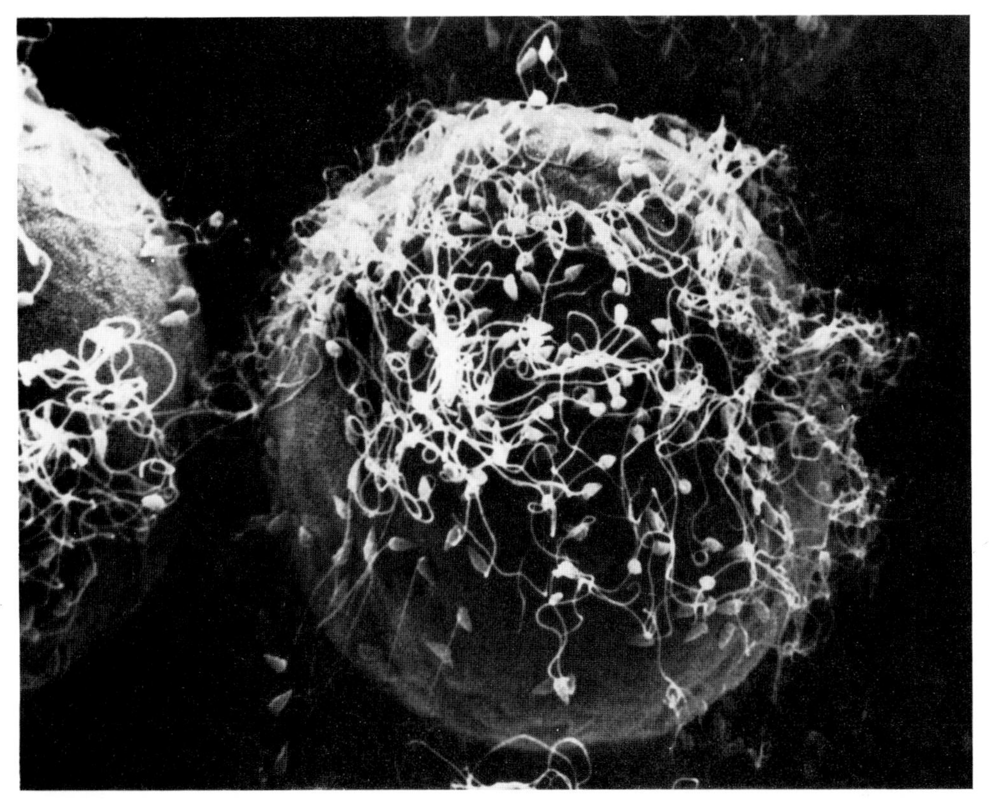

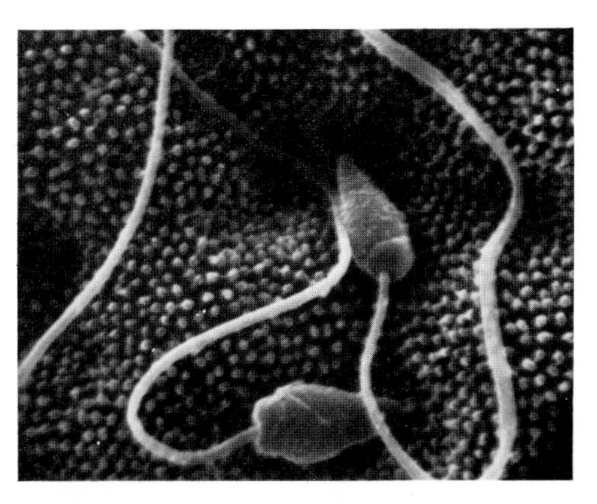

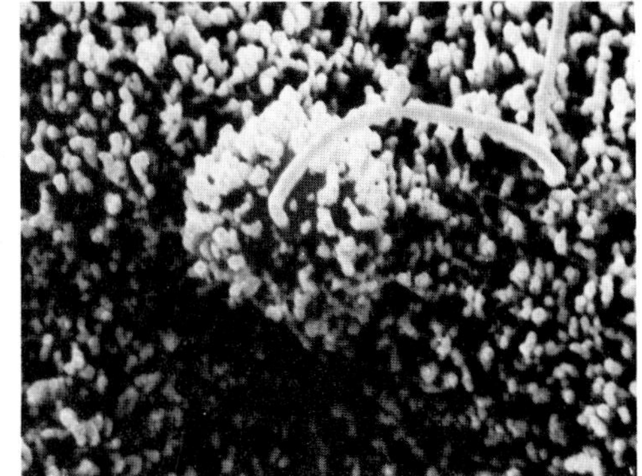

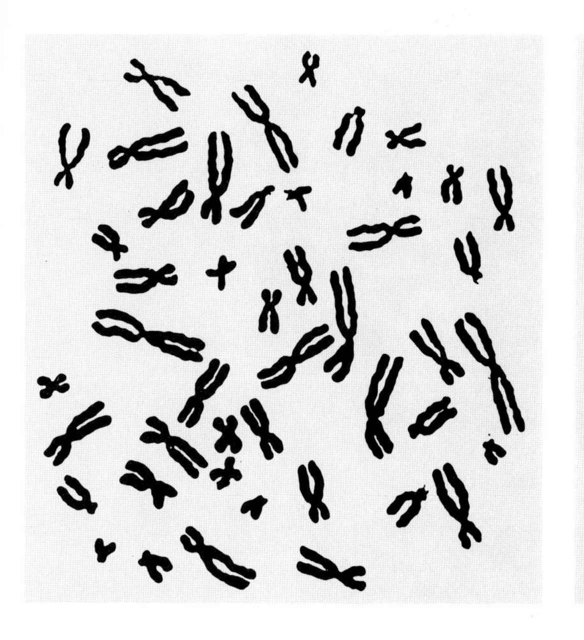

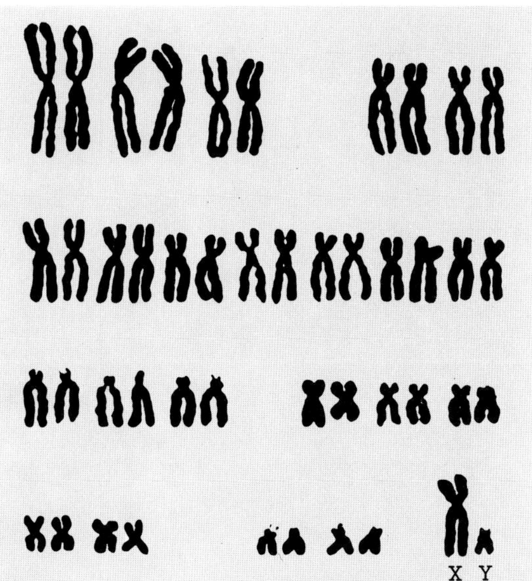

Chromosomes from a human somatic, or body, cell appear scattered at random when viewed under a microscope (far left). However, when the photograph is cut up and rearranged, to show what is termed the karyotype, the forty-six chromosomes form twenty-two pairs, plus two sex chromosomes – one X and one Y. This specimen must therefore be from a male.

sexually and sterile. However, even with a normal complement of sex chromosomes XX or XY, full sexual development also depends upon other factors, for example the level of hormones. Indeed, in some organisms, for example fish, sexual characteristics may be more dependent on environmental conditions during the development of the organism than on which sex chromosomes it has.

Abnormal numbers of other types of chromosomes can also occur in some rare instances. In humans, the fertilized eggs containing these chromosomes often fail to develop very far, but some individuals do occasionally survive and grow to adulthood. Such people frequently have physical and mental abnormalities. For example, individuals possessing one extra (non-sex) chromosome exhibit a condition known as Down's Syndrome (or mongolism). Such individuals have a characteristic facial appearance and limited intelligence.

The diagram below, simplified to show only six of the human chromosomes, shows how meiosis results in the formation of sex cells, which, when fertilized, give rise to either male or female offspring. While both eggs contain the same set of chromosomes, the two sperm differ, one containing an X, the other a Y. Fertilization of an egg by an X-containing sperm results in a female, while fertilization by a Y-containing sperm results in a male.

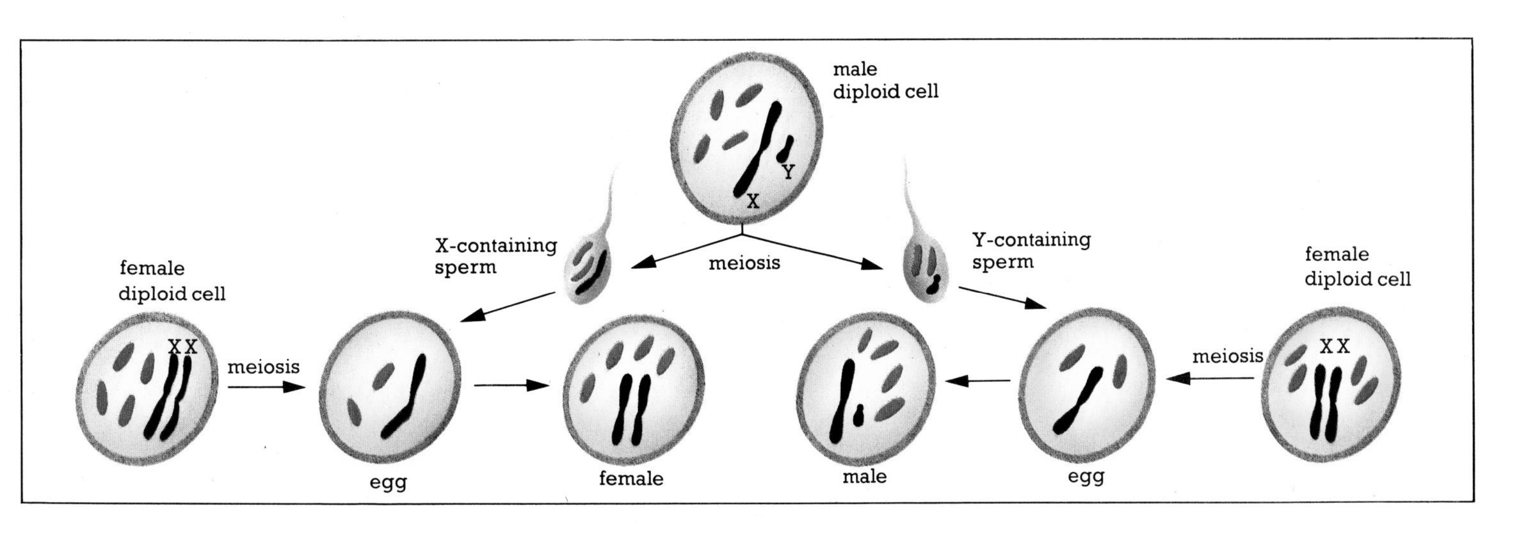

Polyploidy

Occasionally organisms occur that have extra whole sets of chromosomes; that is, instead of the single, haploid set in the sex cells, or the double, diploid set in the body cells (which are usually called somatic cells), they have several complete sets. Such organisms are said to be polyploid. It is fairly common to find groups of related plant species, and some animal species, that have different numbers of chromosomes.

These numbers are all multiples of some basic number. In the family of plants that includes potatoes and tomatoes, the haploid chromosome number is twelve. But species in the family have, for example, 36, 48 and 144 chromosomes. These species are indeed polyploid, having various multiples of the haploid set of twelve chromosomes.

There are several possible ways in which polyploidy may arise. One way is for a plant to have double the normal number of chromosomes in its sex cells, which can happen when both sets mistakenly go to one cell, instead of to two, during meiosis. If such a plant then crossbreeds with a normal plant the offspring will have three chromosome sets. These plants are often sterile, probably because, with an odd number of chromosomes, they have difficulty pairing up and distributing chromosomes during meiosis. However, polyploid plants can often be reproduced asexually by grafting. Some well-known fruits and flowers, for example bananas and certain varieties of tulip, have three sets of chromosomes but can be reproduced asexually.

Polyploidy can also arise if diploid cells fail to distribute the chromosome sets properly during mitosis. This can result in plants with four sets of chromosomes. Again, such plants can often be reproduced asexually and are frequently commercially attractive. For example, tomatoes with four sets of chromosomes have a higher vitamin C content than normal; and some varieties of apple with four sets may have extra large fruits.

Polyploidy can occur among animals, including the Sonoran whiptail lizard, Cnemidophorus sonorae, *shown here.*

Polyploidy can also occur through cross-fertilization between different species. Such offspring would normally probably be sterile. However, if the chromosome number of the offspring accidentally doubles, the organism can then be fertile.

One of the commonest daily foods – bread – is the end product of a number of such events. There are several species of wheat cultivated today for production of different types of flour, or for animal fodder. All wheats are relatives of wild grasses, and the basic, haploid chromosome number in these species is seven. Some 9000 years ago in what is now western Iran, a farmer planted an emmer wheat, a variety similar to that used today for producing flour for macaroni. He happened to plant the wheat in a field near a patch of one of its distant relatives, goat grass, a weed. However, the emmer wheat, a polyploid with twenty-eight chromosomes, was fertilized by the goat grass, an ordinary diploid with fourteen chromosomes, to produce a new polyploid species with forty-two chromosomes. This new species is now the most valuable of all wheats. It is a bread wheat, which plant breeders have developed into over 20,000 cultivated varieties to suit particular soils and climatic conditions.

Emmer wheat (above left) which has twenty-eight chromosomes, and goat grass (above center) which has fourteen, are the probable ancestors of bread wheat (above right), which has forty-two chromosomes.

A number of common foods, such as persimmons, cane sugar, arabica coffee, bananas and bread, come from polyploid plants.

Genetic Mistakes

It is easy to see from the way in which instructions are contained in DNA that any change in the sequence of bases in a gene can change the instructions it contains, and so change the protein corresponding to that gene. This is the chemistry underlying a mutation. The sequence can be changed either by altering bases, or by adding or subtracting bases in the sequence. This new sequence is reproduced and when the cell divides, altered DNA can be passed on to further generations of cells. If any of these cells eventually become eggs or sperm, the mutation can be passed on to future generations of organisms.

Some changes probably occur spontaneously due to small (but potentially serious) mistakes when DNA is being copied; for example, the copying process may put in base C opposite A instead of the usual G. But such mistakes occur only very rarely, about once in every 10,000,000 bases copied.

Certain chemicals, such as mustard gas, and some forms of radiation, such as X-rays, ultraviolet light, and gamma radiation, can increase the frequency of mistakes – that is, the rate of mutation. Many mutations are potentially harmful. If they occur in the genes of eggs or sperm the mutations can be passed on to descendants. It is therefore important to avoid

This diagram shows how a mutation can occur in DNA, shown in the unaltered form at (a). For clarity just one strand of the DNA is shown. An additional base may be inserted (b), or one may be subtracted (c), both causing a whole sequence of bases and its corresponding sequence of amino acids to be read out of phase. Alternatively, an incorrect base may be substituted (d), leading to an alteration in just one amino acid.

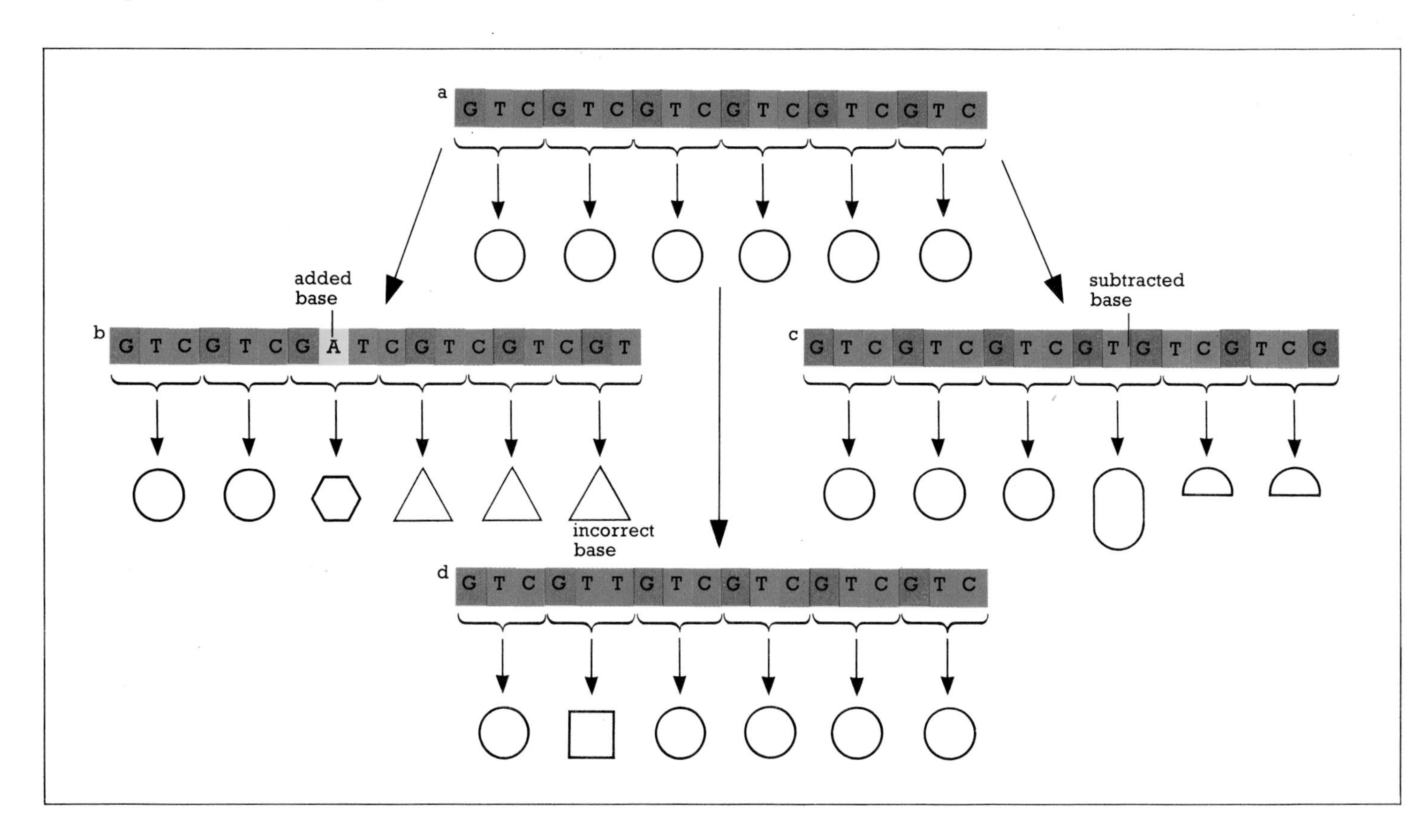

excessive exposure to radiation and other environmental hazards.

Let's consider a mutation that alters a gene so that it codes for an inactive, rather than a normal, active enzyme. Animals and plants usually have two sets of genes. If the copy of the gene inherited from the father is mutant, and the copy inherited from the mother is also mutant, an organism will not have that active enzyme.

The results are sometimes minor: a missing enzyme may merely result in curlier hair, or differently colored petals or eyes. However, in other cases the consequences can be very serious. In humans a number of unpleasant inherited diseases can result from such single missing enzymes. Galactosemia is an example of such a disease, in which one missing enzyme prevents a newborn infant from properly utilizing the carbohydrate in its food.

But even if a cell makes a mistake in the sequence of bases in its DNA, or has an error forced on it – say by radiation – this does not necessarily lead to a mutation. Cells have a number of special enzymes that can examine the sequence of bases in DNA and spot errors. Such enzymes work rather like proofreaders who read a book before it is finally printed to check that it contains no mistakes. If an error, such as a misplaced base, is detected, the enzyme can remove the offending base and replace it with the correct one. However, the cell's proofreading enzymes sometimes fail to notice, and correct, an error in the DNA before it is reproduced and passed on. In this case a mutation will still slip through.

Occasionally a proofreading enzyme can itself be abnormal because its own gene is mutated. In humans one such abnormality can lead to a form of skin cancer. Most people who like to get a nice suntan are benefiting from the sun's ultraviolet rays. But ultraviolet rays can cause changes in the bases in DNA. Normal people do not seem to be affected by this, because the cell's proofreading enzyme can cope with such changes. People lacking a normal proofreading enzyme are heavily freckled and long exposure to sunlight can ultimately lead to skin cancer. Fortunately this severe condition is recessive and occurs only in people inheriting the gene for abnormal proofreading enzyme from both parents. Where one normal gene is inherited, the people are freckled but otherwise healthy.

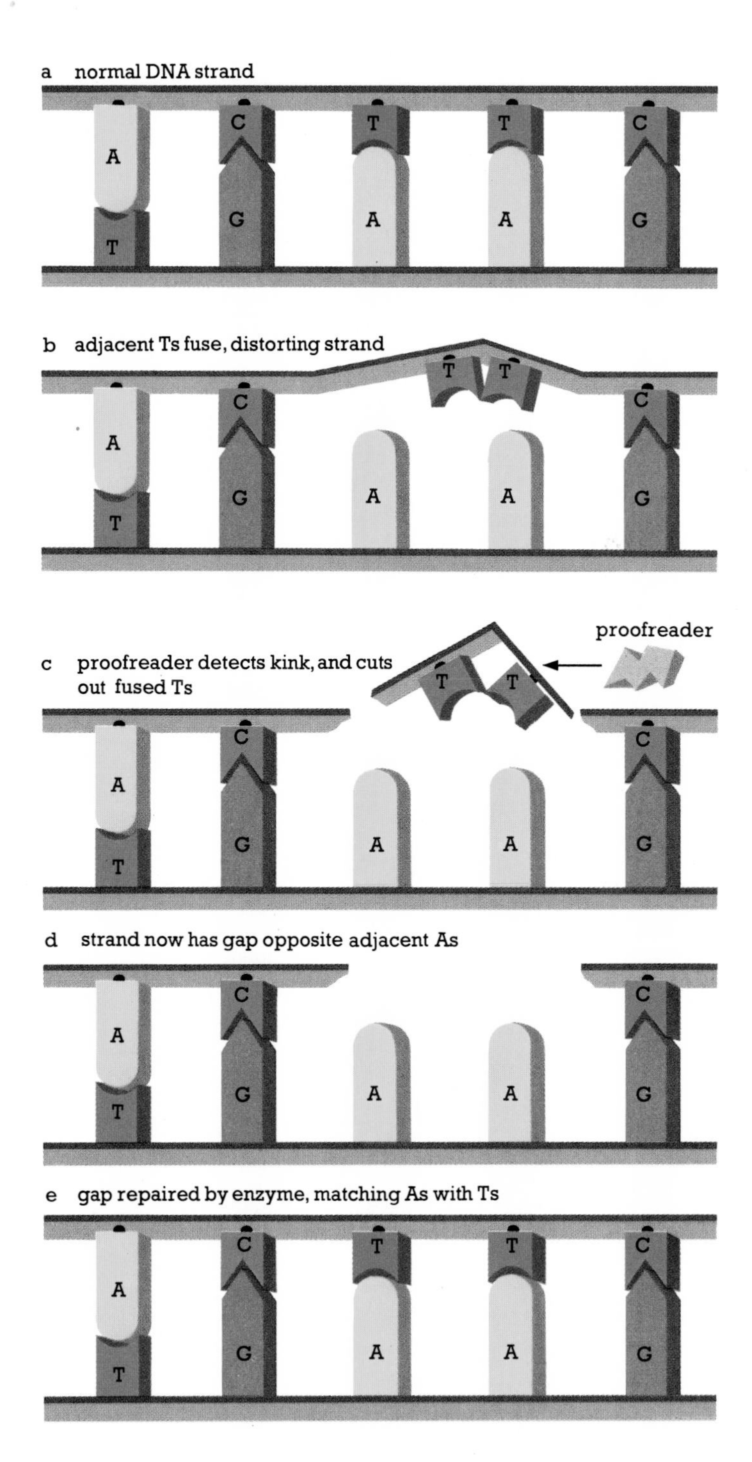

Ultraviolet light can cause two adjacent thymine (T) bases (blue) in a normal DNA strand (a) to fuse together (b). This can lead to a mutation unless the enzyme that checks for mistakes (the proofreader) detects the distortion. The enzyme cuts out the linked thymines (c–d) and replaces them with two new bases (e), thus restoring the DNA to its original form.

Mendel and His Peas

In the middle of the last century, Gregor Mendel, a monk who lived and taught in a monastery in Brno, in what is now Czechoslovakia, decided to investigate the mechanism of inheritance in plants. He worked on edible garden pea plants, which he could breed easily and grow in his monastery garden. Rather than just examine plants to see how much they resembled their ancestors in some vague general way, he wisely decided to follow carefully the inheritance of just a few specific characteristics. He chose plants showing clear-cut differences, such as the height of the plants and the shape and color of the pods and the peas. Mendel's aim was to discover what controlled such characteristics, and how they were inherited.

Each flower of a pea plant contains both male and female sex cells. If left alone, these will fertilize themselves and will eventually produce seeds (the peas that are eaten). When planted, these seeds will grow into adult plants. But by taking pollen from one plant and brushing it on to the female parts of another plant Mendel was able to achieve cross-fertilization. He called the offspring that resulted from such crosses hybrids.

Mendel was able to produce hybrid offspring for several characteristics. For example, he crossed a tall plant with a dwarf one, and a plant grown from green seeds with one grown from yellow seeds. In each case he found that the hybrid offspring all resembled one parent rather than the other. In the experiment on plant height, for example, all the hybrids obtained from crossing tall and dwarf plants were tall. It appeared that whenever alternative states existed for a characteristic, one state was dominant over the other, recessive state. However, when the hybrids were allowed to self-fertilize, some of the "grandchildren" of the original cross resembled the dominant "grandparent" while some resembled the recessive one. This suggested to

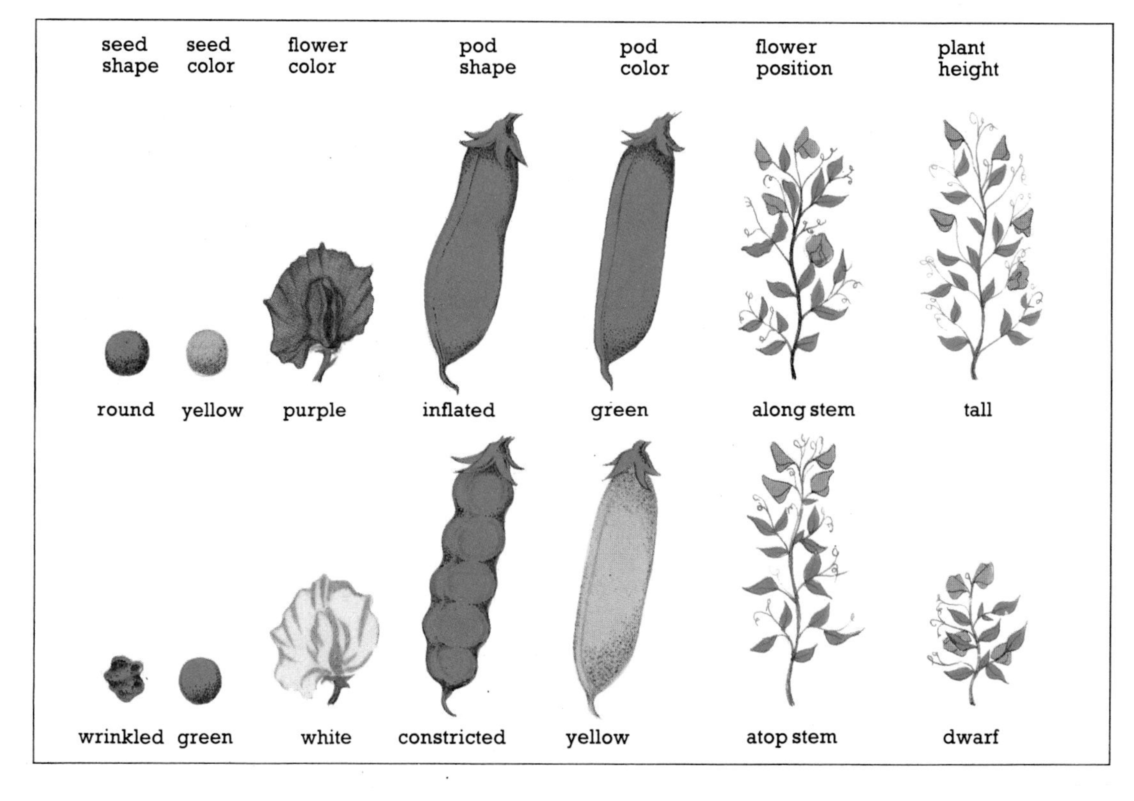

Mendel studied pea plants in order to discover how certain characteristics are inherited. The dominant forms of the seven characteristics with which he experimented are shown here in the top row, and their recessive counterparts are shown below.

Mendel that whatever was responsible for the inheritance of the recessive state must have been present in the first-generation hybrids; however, its effect was masked by whatever determined the alternative, dominant state.

Mendel was not the first plant-breeder to notice these phenomena of dominance and the reappearance of recessives, but he was probably the first to appreciate their full significance. This was partly due to something he did that, strange as it now seems to us, was very rare among nineteenth-century biologists: he counted his plants. When he calculated the proportions of dominants and recessives he was struck by a remarkable regularity. When the hybrids were allowed to self-fertilize, Mendel found that for every offspring that showed the recessive state of the characteristic, three showed the dominant state. What could this recurring three to one ratio mean?

With amazing insight, Mendel realized that if at fertilization each parent contributed one sex cell containing one factor responsible for each characteristic, then the results could be explained. He suggested that each factor (now called a gene) could exist in two alternative forms. One form could contain instructions for a dominant state of a characteristic, the other could contain instructions for a recessive state. If the plant inherited either two genes in the dominant form, or one dominant and one recessive, then it would show just the dominant state. Only if both genes were in the recessive form would the plant exhibit the recessive state. On the average, three-quarters of the offspring of a self-fertilized hybrid would inherit one or both genes in the dominant form and hence exhibit the dominant state; while one quarter would inherit both genes in the recessive form and exhibit the recessive state; hence the three quarters to one quarter, that is the three to one ratio.

Even today useful knowledge can still be gained from carefully controlled breeding experiments – essentially what Mendel did. Mendel's Laws, derived from peas, seem to hold true for most plants and animals. Thus the ratio of particular alternative states of a characteristic can give useful clues as to the number of genes involved in its inheritance. For example, a ratio other than three to one for a characteristic among the "grandchildren" of a cross-fertilization may suggest that more than just one gene is involved in the inheritance of the characteristic.

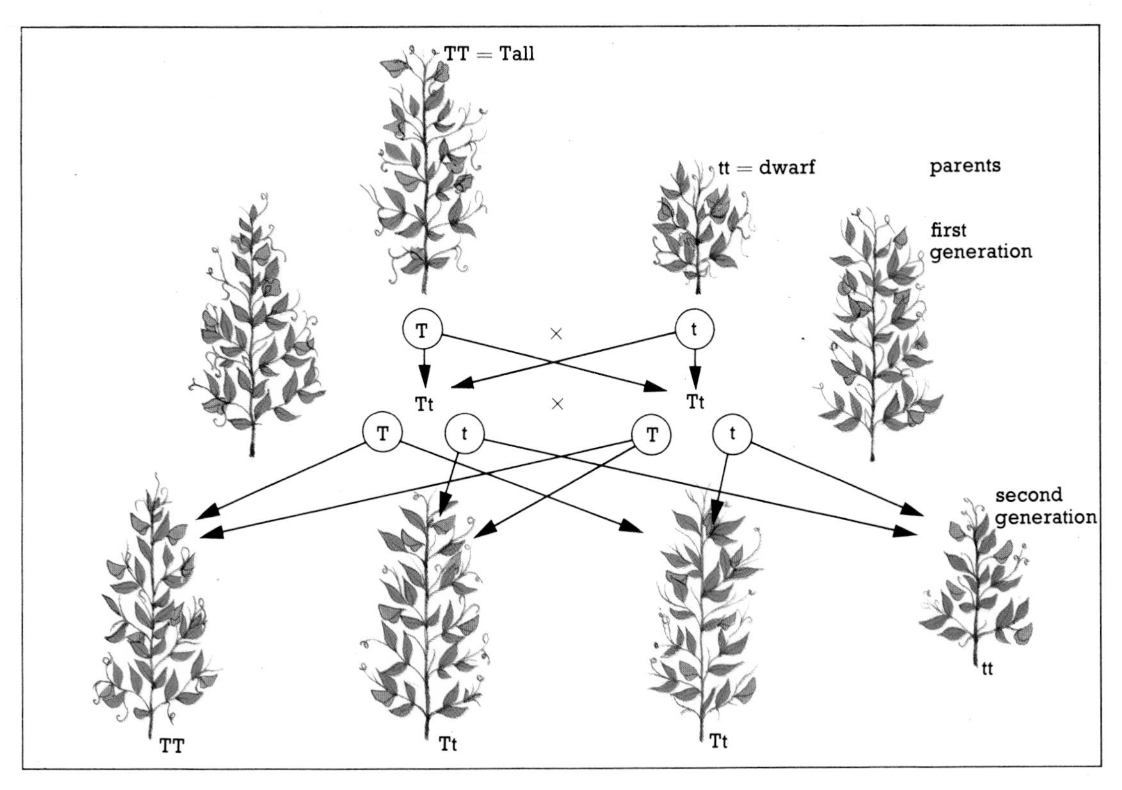

This diagram indicates the results Mendel obtained when crossing two varieties of pea plant. By convention, capital letters designate the dominant form (tall), while small letters designate the recessive form of the gene (dwarf). In the first generation the offspring are tall, but in the second generation one in four plants are dwarf. The actual ratio Mendel obtained was 787 tall to 277 dwarf, or 2.84 to 1.

Mapping Genes

Geneticists often speak of an organism inheriting a specific gene, or a number of specific genes. Mendel's method of studying inheritance allows researchers to follow the inheritance of genes over many generations. Organisms actually inherit a large number of genes, but it is much easier to study just a few at a time.

An animal or plant inherits its genes on a number of chromosomes and each chromosome contains hundreds or thousands of genes. If two genes happen to occur on different chromosomes, then they will be inherited independently. On the other hand, if two genes happen to occur on the same chromosome, then they will tend to be inherited together: their inheritance will be linked. So, in breeding experiments, genes that occur on the same chromosome might be expected to stay together generation after generation.

There is one complicating factor. During meiosis chromosomes can break, cross over, and rejoin. When this happens, the free end of a chromosome derived from the mother may join up with one derived from the father, and vice versa. The chromosomes derived from the mother and the father, although broadly similar, will probably differ in their detailed instructions (for example, brown eyes or blue eyes in a gene for eye color, red or black hair in a gene for hair color). The crossing over of chromosomes can therefore lead to new combinations of genes in some sex cells, and hence in some offspring. An experimental breeding test shows that for any two linked genes, a majority of the offspring will contain the same combinations of characteristics found in the parents. A small proportion of the offspring will carry the new combinations.

The actual proportions of offspring with the new combinations of characteristics provide clues about relative positions of genes along the chromosomes.

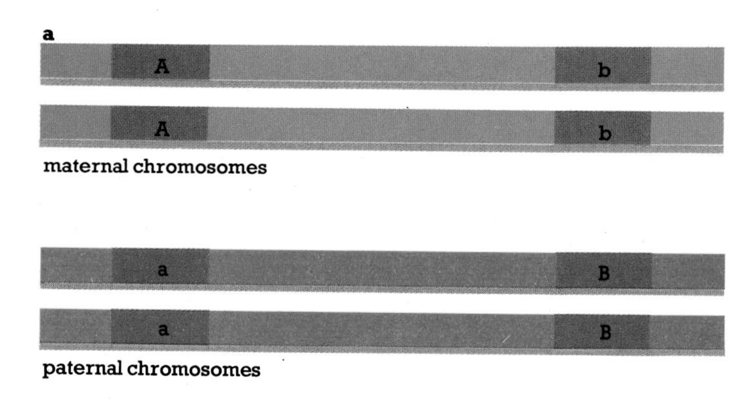

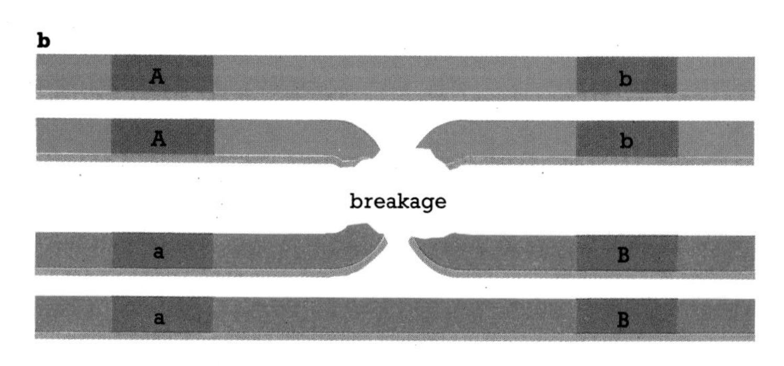

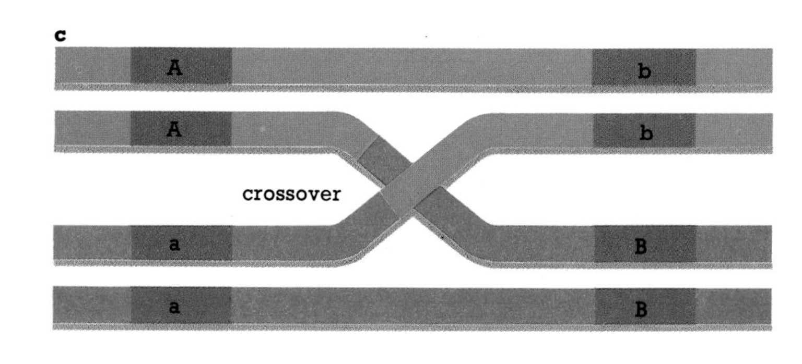

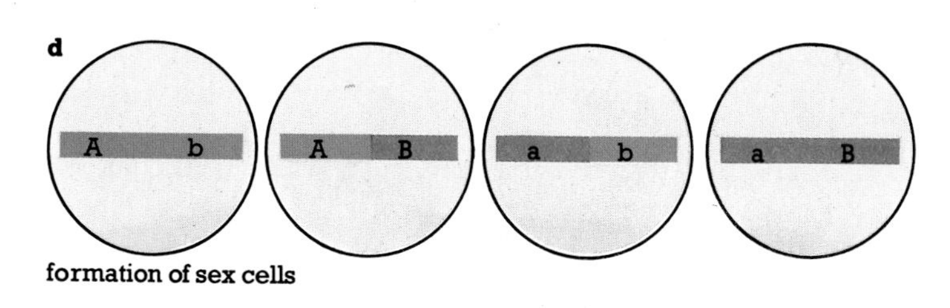

This diagram shows how maternal and paternal chromosomes (a) can undergo breakage (b), and rejoin (c) to form new combinations of genes in two of the four sex cells after meiosis (d).

The closer together two genes are located on the chromosome, the less likely it is that a break will occur between them during meiosis.

Picture the chromosome as a clothesline, with a sequence of colored clothespins for the individual genes. Assume that the sequence is: red – green – blue – white – black – orange – purple. If you took a large sharp knife, closed your eyes, and slashed at the line at random, you would be less likely to cut the line between pins that were close together (like white and black) than between pins that were far apart (like red and black). In the same way, the chances of a break in a chromosome occurring between closely linked genes are much smaller than for distantly linked genes.

By counting the proportion of offspring where particular characteristics have remained together or have been separated during inheritance, geneticists can gain some idea about the location of genes in relation to each other on a chromosome and can represent this as a gene map. Just as a map of houses along a street might give an idea of the relative positions of Mr. Jones at number 170 to Mrs. Smith at number 187, and Dr. Brown at number 223, so a gene map shows the relative position of genes along the chromosome. For an organism with several chromosomes, each containing its own distinctive set of genes, geneticists can construct a separate map for each chromosome.

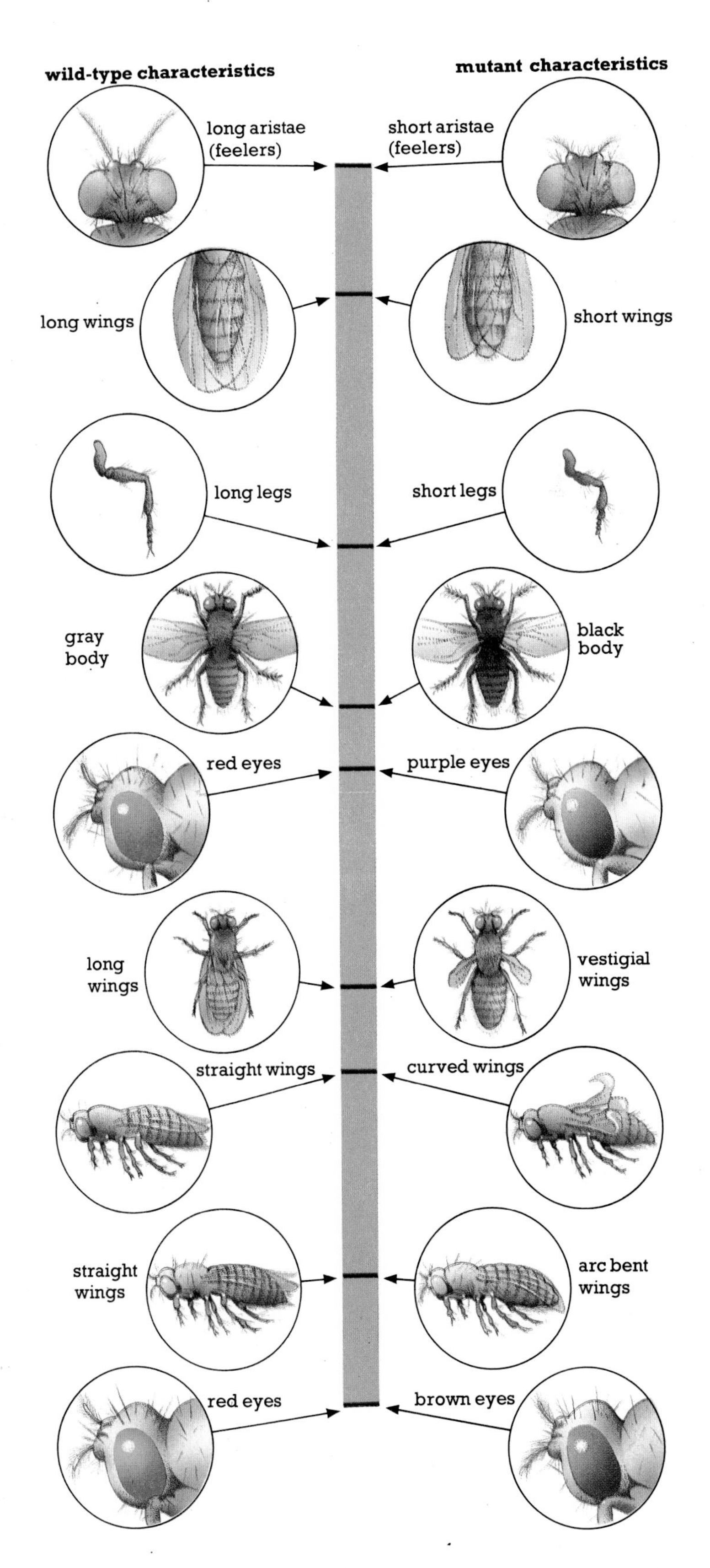

A simplified gene map shows the relative positions of certain genes on part of one of the four chromosomes of a fruit fly. Geneticists identify genes by the wild-type and mutant forms shown. Varied distances between genes indicate greater or lesser chances of crossover, and of new combinations of genes in an offspring. For example, there is a greater chance of crossing over between genes for long feelers and long legs than between the gene for a gray body and the one nearby for red eyes.

Sex Linkage

Genes that occur on sex chromosomes – and the corresponding characteristics – are said to be sex linked. This relationship can lead to some interesting patterns of inheritance that differ between males and females.

As females have two X-chromosomes, a recessive mutation that occurs on one of these X-chromosomes is likely to be offset by a dominant normal wild-type gene on the other X-chromosome. Males, on the other hand, have only one X-chromosome along with a Y-chromosome. If a recessive mutation occurs on the X-chromosome, there is no offsetting dominant wild-type gene, and the mutation will appear.

The result can take some interesting forms. For example, there is a mutant gene in the fruit fly leading to white eyes instead of the wild-type red. This mutation generally only appears in male flies because the gene governing eye color is on the X-chromosome. A female fly will almost always have red eyes since mutations are rare, and the likelihood is that a female fly's second X-chromosome will have a wild-type gene. This will dominate any recessive mutant gene on the other X-chromosome.

When the mutant gene is carried by a female and passed on by her, its effect may be felt in some of her sons, not in her or her daughters. By contrast, a male may show the abnormal characteristic but cannot pass the abnormal gene on to his sons. This is because sons receive their Y-chromosome from their fathers, and their X-chromosome from their mothers.

By a rare coincidence it is possible, however, for a white-eyed female fly to occur. If a white-eyed male (bearing the rare mutant gene) mates with a red-eyed female, in which one X-chromosome happens to have the rare mutant gene, then on the average half of the daughters will be red-eyed and half white, depending on which X-chromosome they receive from their mother.

Little is known about genes that occur on the Y-chromosome. In humans only one characteristic is known that probably results from a mutation in such a gene. This mutation results in hairy ears.

There are, however, a number of human characteristics that result from a mutant gene on the X-chromosome. One form of color blindness, where the individual cannot distinguish between green and red, is sex linked. As would be expected, it occurs much more frequently in men than in women, since men cannot have a second compensatory X-chromosome. As with white eyes in fruit flies, the mutant gene is carried by women and can be passed on to their sons.

There is another sex-linked characteristic that has had a very interesting history. This is the rare and often fatal disease called hemophilia, in which the blood fails to clot when a wound occurs. The gene responsible is contained on the X-chromosome, so many more men have the disease than women. But, of

This diagram shows how a female fruit fly with wild-type red eyes can still act as the carrier of a white-eyed mutation (marked with a black cross) on one of her two X-chromosomes. This recessive mutation will be passed on to an average of fifty per cent of her male offspring.

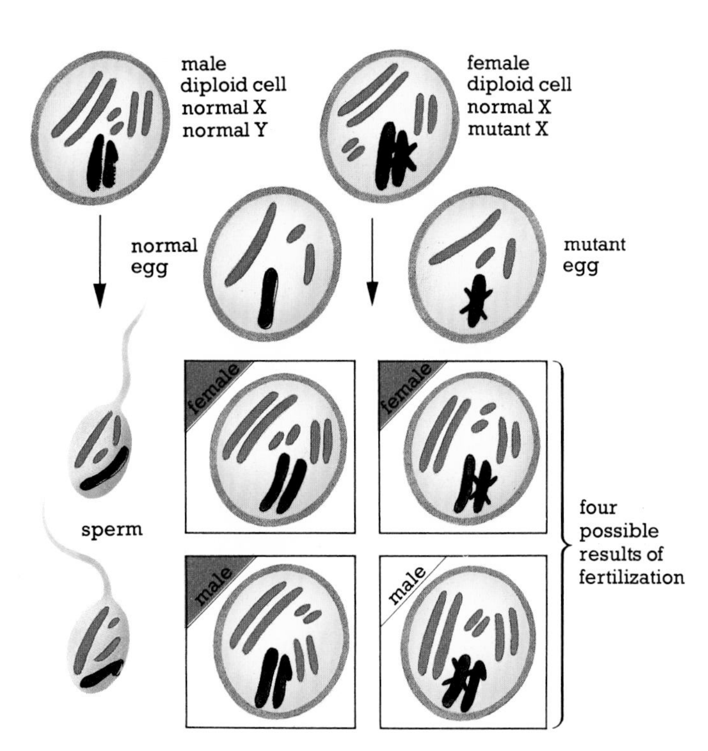

course, normal women who carry the rare mutant gene on one of their X-chromosomes can pass it on to their sons.

Britain's Queen Victoria seems to have been one such woman. She had nine children, of whom one son was a hemophiliac and at least two daughters were carriers. As a result, hemophilia has periodically cropped up in male members of the European royal families into which they married. The results are therefore well documented.

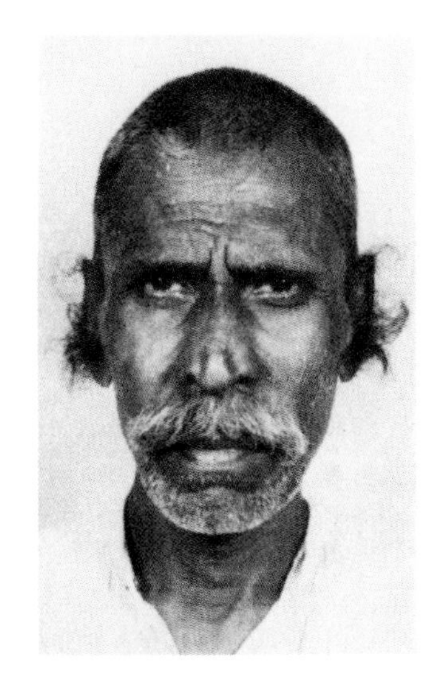

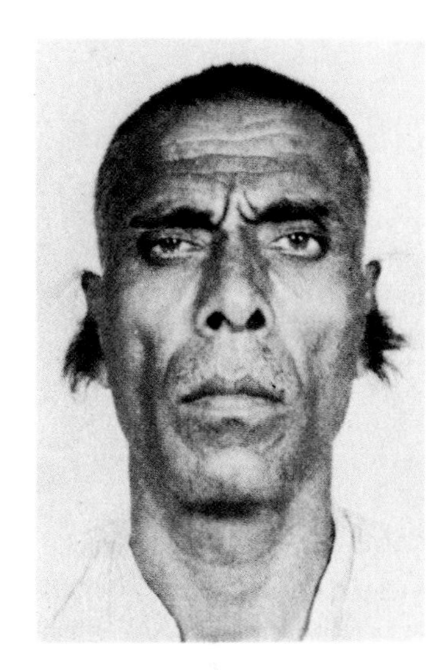

The hairy ears of these two brothers from India are a characteristic resulting from a mutation on a gene present on the Y-chromosome.

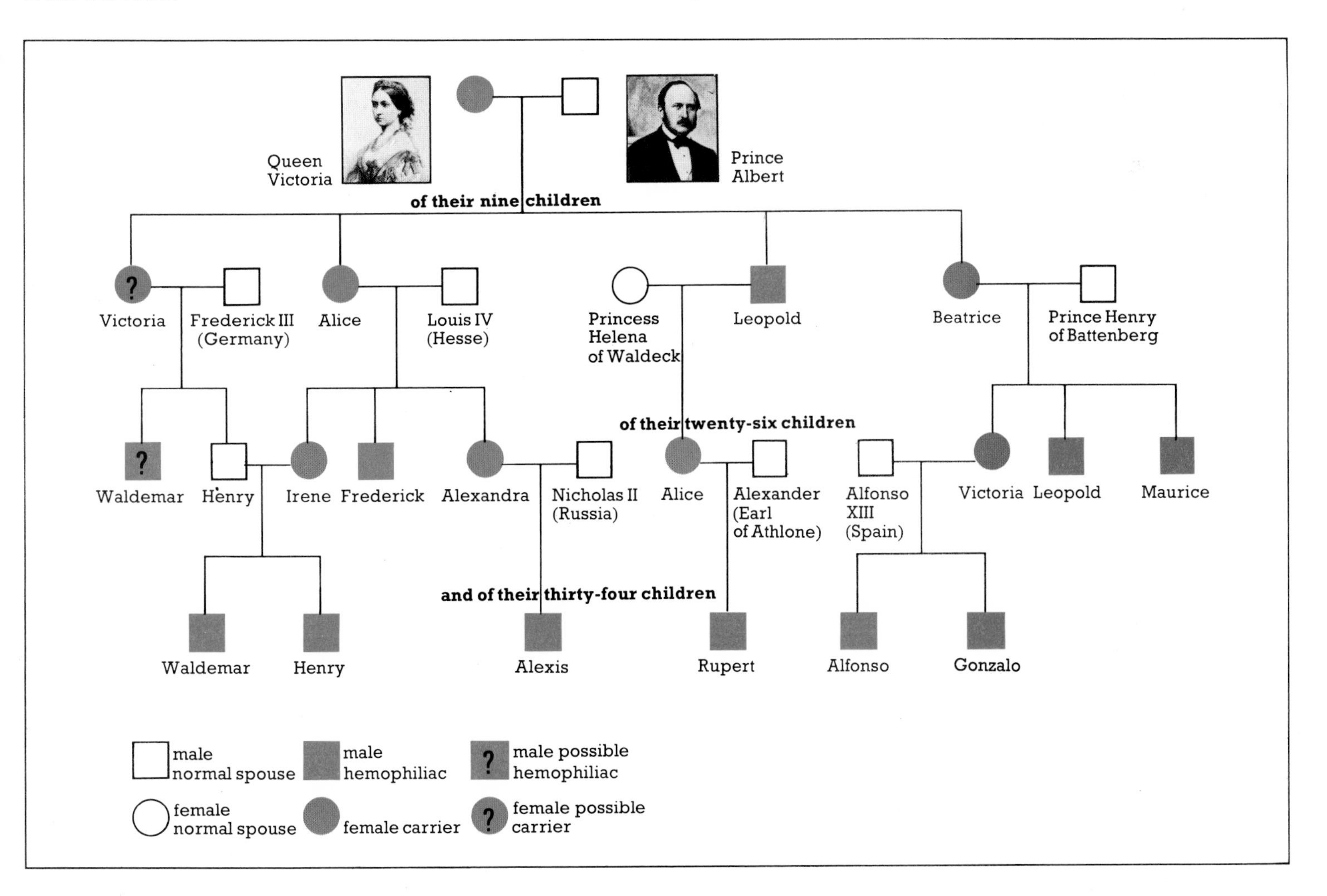

This simplified family tree shows the inherited tendency of hemophilia in the descendants of Queen Victoria and Prince Albert. The disease is only evident in males, many of whom died relatively young, yet its occurrence reveals that females, including the Queen herself, acted as the carriers of the disease.

Gene Interaction

Ultimately the characteristics of a cell, and of the organism as a whole, depend on what enzymes and structural proteins it possesses, and therefore which chemicals and structures it makes. One enzyme rarely acts alone, and the production of a chemical substance frequently depends on several enzymes acting in sequence. Since the formation of each enzyme is determined by a particular gene, the precise appearance or functioning of an inherited characteristic often results from a combination of genes rather than a single gene. For example, the precise shape of your nose may depend upon the interaction of various enzymes and structural proteins; this in turn depends on the exact combination of genes you have inherited that are concerned with the nose.

Since combinations of genes occur in the offspring that are not present in either parent, some characteristics of the offspring can be different from those of either parent. For this reason, the pollination of a yellow flower with pollen from another yellow flower could, in principle, produce seeds that, when planted, grow into red flowers. Similarly, chickens can occur with combs quite unlike those in the parents or the grandparents. It is also possible for a particular combination of genes to reappear after being absent for several generations and this can result in a throwback – an individual that resembles its distant ancestors, rather than its parents or grandparents.

Many characteristics of living organisms are essentially a matter of degree. For example, animals and plants are not simply tall or short, heavy or light; instead, a wide variety of intermediate sizes and weights is found. Similarly, in human beings there is a wide range of skin colors between the extremes of black and white. Characteristics like these probably depend upon combinations of numbers of genes, each of which helps determine the final appearance. For example, if an organism inherits a high proportion of genes all contributing towards dark skin or

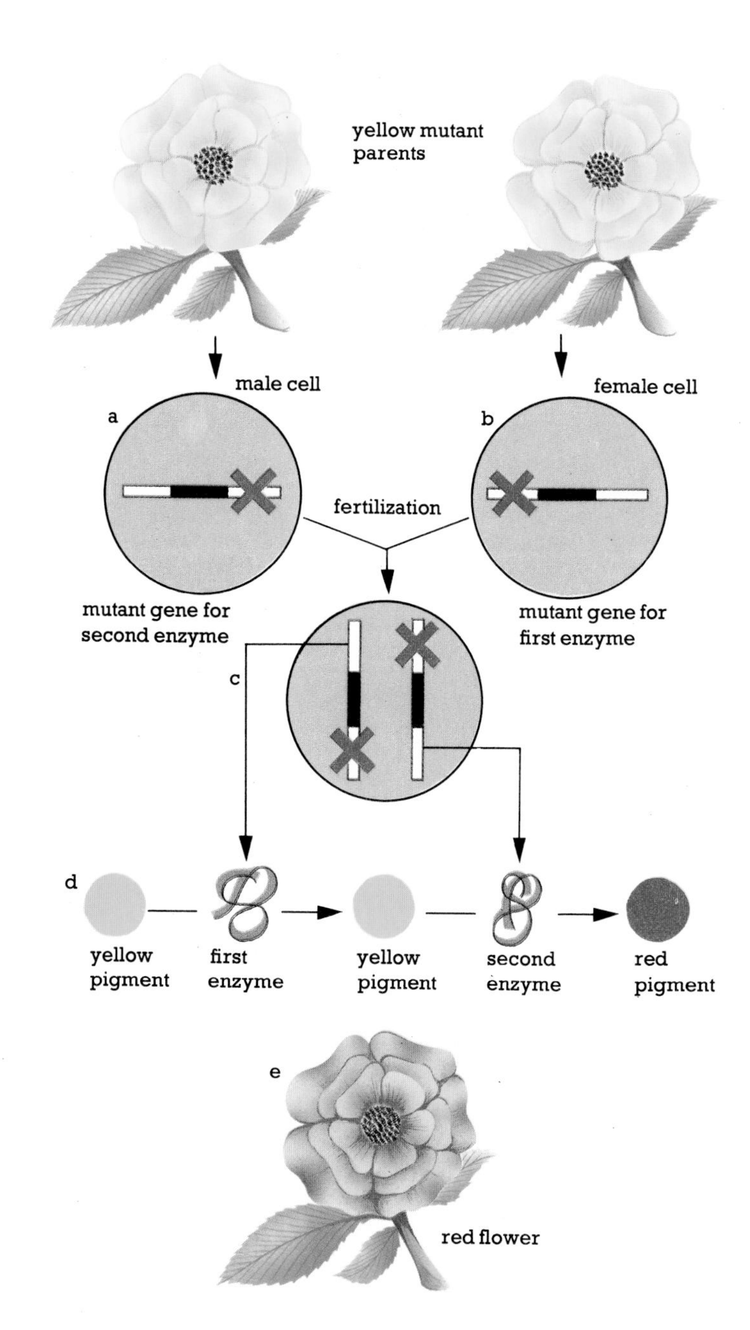

In the diagram above the yellow flowers (top) are mutant forms of a red variety that normally produces red pigment from yellow by a pathway requiring two enzymes. Each plant lacks one of the enzymes (a and b). Their offspring inherit a normal copy of each gene (c) and so can make both enzymes, completing the pathway to make red pigment (d) and producing a red flower (e).

a

b

c

long petals, the organism will be very dark or have very long petals. On the other hand, if only a few of the genes are in the form leading to dark skin or long petals, the organism will be relatively light in color, or have petals of moderate length.

It is very difficult to tell how many genes are involved in any such graduated characteristics. Geneticists use complex methods of statistical analysis but can only suggest the likely number. In addition, the role of environment may be considerable. The height a plant grows to may well depend not only on what precise combinations of genes it has inherited, but also on the altitude and climate at which it grows, the soil, and the food it receives.

The apparently wild pigeon (a) can crop up as the offspring of two specially bred pedigree parents (b and c). This throwback has probably arisen because of a combination of recessive genes that has not occurred for several generations.

In the chickens shown below the four types of comb walnut or strawberry (a), pea (b), rose (c) and single (d) – are the result of the interaction of two genes. Each of these genes can exist in two alternative states – one dominant and one recessive.

a

b

c

d

Genes and Environment

In a game of billiards between two people, the players use only red and white balls and enjoy a good, evenly matched game. Then they switch to a game with a variety of colored balls, such as snooker or pool, and one person starts to behave rather strangely. He hits brown or green instead of red, and generally seems a bit puzzled.

What has gone wrong? One possibility is that he is color-blind and cannot distinguish red from brown or green. If so, he may have inherited this visual problem because of a mutation on his X-chromosome. However, his behavior only became strange when he started playing snooker or pool. In other words, his abnormality depended not only on his genes, but also on the environment with which they interacted. In one environment (playing billiards) he is normal, in others (playing snooker or pool) his genes put him at a disadvantage. All genes interact with the environment and the effects of some, like those for color blindness, are strongly dependent on it.

In other cases, the effects of certain genes can be altered by changes in environment. For example, someone who is near-sighted because of a hereditary defect can read just as well as a normal sighted person if provided with the "environment" of suitable spectacles; the genes remain unaltered but their effect is greatly modified.

Organisms with identical genes can also appear or act very differently if placed in different environments. For example, plants with identical genes will grow quite differently if placed in different types of soil or climate.

Even characteristics that are apparently rigidly inherited may vary in different environments. Himalayan rabbits have pink eyes and distinctive black patches – characteristics that are passed on from generation to generation. However, if we pluck fur from the white part and keep the bald patch cold, the fur will regrow black, not white. On the other hand, if we pluck the black hair and keep the bald patch warm, it will regrow white.

What Himalayan rabbits seem to inherit, therefore, is not the pattern of black and white fur as such, but a tendency for certain areas of skin to grow

In ordinary weather conditions (a), a road user with normal vision (left) relies on the color of traffic lights. The color-blind road user (right) relies on their position: top means stop, bottom means go. In a fog (b), a color-blind person is at risk as he is unable to distinguish between the color of the lights or their position. Better-designed lights, such as flashing green, could remove this problem, which is a direct consequence of a mutant gene.

By shaving a patch of fur from the back of a Himalayan rabbit (a) and keeping the patch cool with an ice pack (b) scientists produce a dark patch of fur (c). This experiment shows that the genes for light hair are capable of producing dark hair, given a modified environment.

black or white hair, depending upon temperature.

One of the most important environmental influences for a developing organism is the food it eats. In a hereditary disease of humans, known as phenylketonuria, an amino acid found in many foods is not properly converted by the body. This is due to a mutation in a gene that contains the instructions for an enzyme involved in the conversion of the amino acid. Failure to convert the amino acid correctly can lead to very severe symptoms: an affected child fails to develop normally and may, as a result, become incurably retarded mentally. However, if the disease is detected early enough – for example, by a blood test – and the baby is fed a diet from which much of the troublesome amino acid is removed, development returns closer to normal.

This example shows how a single inherited gene mutation can, in one set of circumstances, produce severe defects. Yet a simple change in the environment – in this case, a baby's diet – compensates for the mutation and largely nullifies its effect.

All genes act within, and interact with, the environment. To ask how much of a characteristic is due to the genes and how much is due to the environment is meaningless. A more useful question is how a particular gene, or genes, will respond to certain environments. The answer may make it possible to manipulate the environment advantageously.

When a variety of Potentilla glandulosa *is grown at different altitudes the response can differ widely. This particular variety grows best at* 4500 *feet (far left), the altitude at which it occurs normally. At both sea level (center) and at 10,000 feet (left), growth is more feeble.*

Immune Systems and Blood Groups

All organisms have some protection against foreign intruders. Animals, including humans, have a sophisticated series of mechanisms called the immune system. The immune system reacts to intruders because they contain chemical substances that are normally absent from the body. To resist these foreign substances, cells in the blood first produce a series of special proteins, called antibodies. Each foreign substance triggers off the production of a specific type of antibody, whose molecules attach themselves to those of the intruder. The antibody acts as a sort of label that enables the foreign invader to be recognized and destroyed by amoeba-like white cells that are always present in the bloodstream.

The first time the body meets any particular type of foreign substance it produces a relatively small amount of antibody to counterattack. In subsequent encounters the immune system is ready to attack more vigorously. This is why people are frequently more resistant to a disease if they have already had it once, and explains why it is possible to increase resistance to disease by artificial means. This is done by giving them injections of disease-causing viruses or bacteria, which have been treated to make them harmless, but still capable of stimulating the body's immune system. Such inoculations marshal antibodies against diseases such as smallpox, cholera, and measles.

An animal's immune system can neutralize a great variety of foreign substances, but in some circumstances the immune system's ability to distinguish friend from foe can be a handicap, as in blood transfusions.

Blood transfusions are used to replace blood lost by injury or disease. Human blood has a number of components, but the most significant ones in transfusions are the red blood cells, which carry oxygen, and those components that form the fluid called blood plasma, in which the red cells are suspended.

Blood is classified into four groups according to the types of molecules found on red blood cells. Because blood groups depend on genes the possible blood groups of children (center column) can be predicted once the blood groups of their parents (left-hand column) are known. Equally, the occurrence of certain blood groups among the children can be ruled out (right-hand column).

parents' blood types	possible children's blood types	impossible children's blood types
A + A	A, O	AB, B
A + B	A, B, AB, O	
A + AB	A, B, AB	O
A + O	A, O	AB, B
B + B	B, O	A, AB
B + AB	A, B, AB	O
B + O	B, O	A, AB
AB + AB	A, B, AB	O
AB + O	A, B	AB, O
O + O	O	A, B, AB

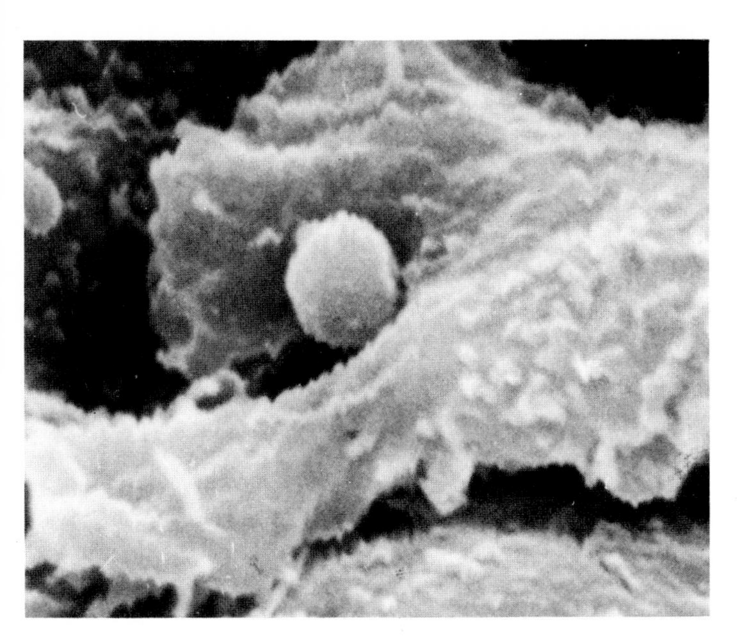

This large, irregularly shaped, amoeba-like cell is found in the lungs. It is seen here in pursuit of the foreign particle in the center of the photograph, which it will destroy by digesting it. Cells similar to this one are a vital part of the body's defense against disease.

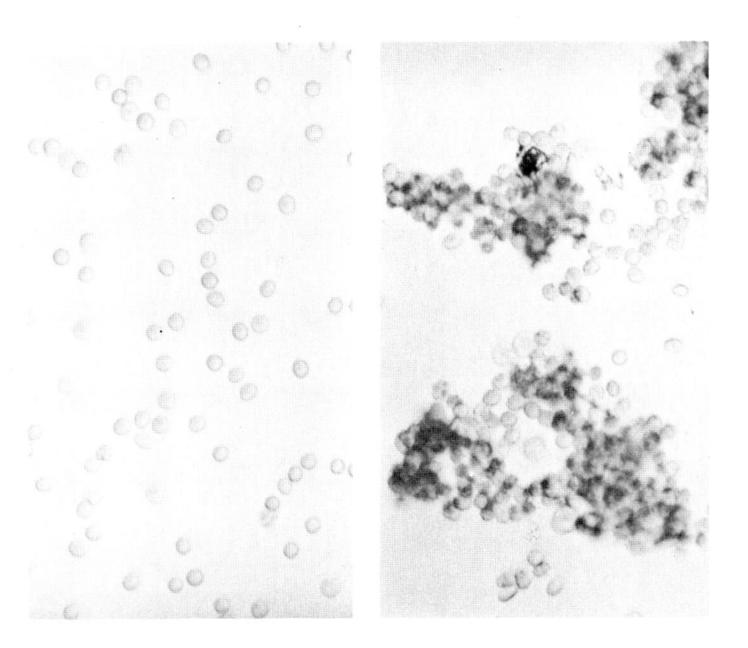

These photographs (magnification $\times$600) show how doctors can determine a person's blood group by adding antibodies against specific blood groups to a sample of the blood (above left). Blood that is incompatible appears clotted when viewed under a microscope (above right).

Specific types of molecules on the surface of the red blood cells enable blood to be classified into four groups. If the cells carry type A, the person is of blood group A; if type B, they are of blood group B; if they have both types, they are of blood group AB; and if they have no such types, they are of blood group O. The types that a person has depend on three forms of a single gene. One form (GA) specifies the production of type A; another form (GB) specifies production of type B; and the third form (GO) does not specify any type. Human chromosomes occur in pairs and so, of course, a cell has two copies of the gene. GA and GB are not dominant or recessive in relation to each other, so a person with both forms of the gene will also have both types of molecules, A and B, on their red blood cells. Therefore, the blood group of this person is AB.

Blood plasma contains antibodies against alien blood groups. For example, the plasma of a person with blood group A contains antibodies against type B. These antibodies are always present. They do not depend for their production on a previous encounter with an alien blood group. This factor complicates blood transfusion.

If type A blood is given to a person with type B, for example, the antibodies present in the patient's bloodstream against type A will attach themselves to the incoming red blood cells, and cause them to clump together. The action of antibodies against incompatible red blood cells can lead to a whole host of effects – notably shock – that can kill the patient. It is therefore vitally important to know the blood group of a potential donor and recipient before transfusion, to ensure that they are compatible.

Fortunately, the clumping reaction caused by antibodies in plasma can be produced in the laboratory, and this enables doctors to determine blood group very easily. Nowadays large blood banks are kept, where all the donated blood has been carefully classified and is available for any patient needing blood of the same group.

The Rhesus Factor

There are many genetically determined factors distinguishing blood groups in addition to the ABO system. The most important of these is the rhesus system because it can cause complications during pregnancy. The rhesus (or Rh) factor owes its name to its initial discovery in the blood cells of rhesus monkeys. Most people have the factor and are called Rh-positive, while those without it are called Rh-negative.

Mother and fetus will sometimes be of different ABO blood groups, but normally the fetus is well protected from the immune system of its mother by the placenta, a membrane barrier that allows nourishment to pass from the mother's bloodstream to the fetus, and waste products from the fetus's bloodstream to the mother's. The placenta also acts as a barrier to the mixing of fetal and maternal blood cells. However, if the Rh factor genes differ in a crucial way between mother and fetus, serious complications can arise, which may lead to the death of the fetus.

The gene for the presence of the Rh factor is

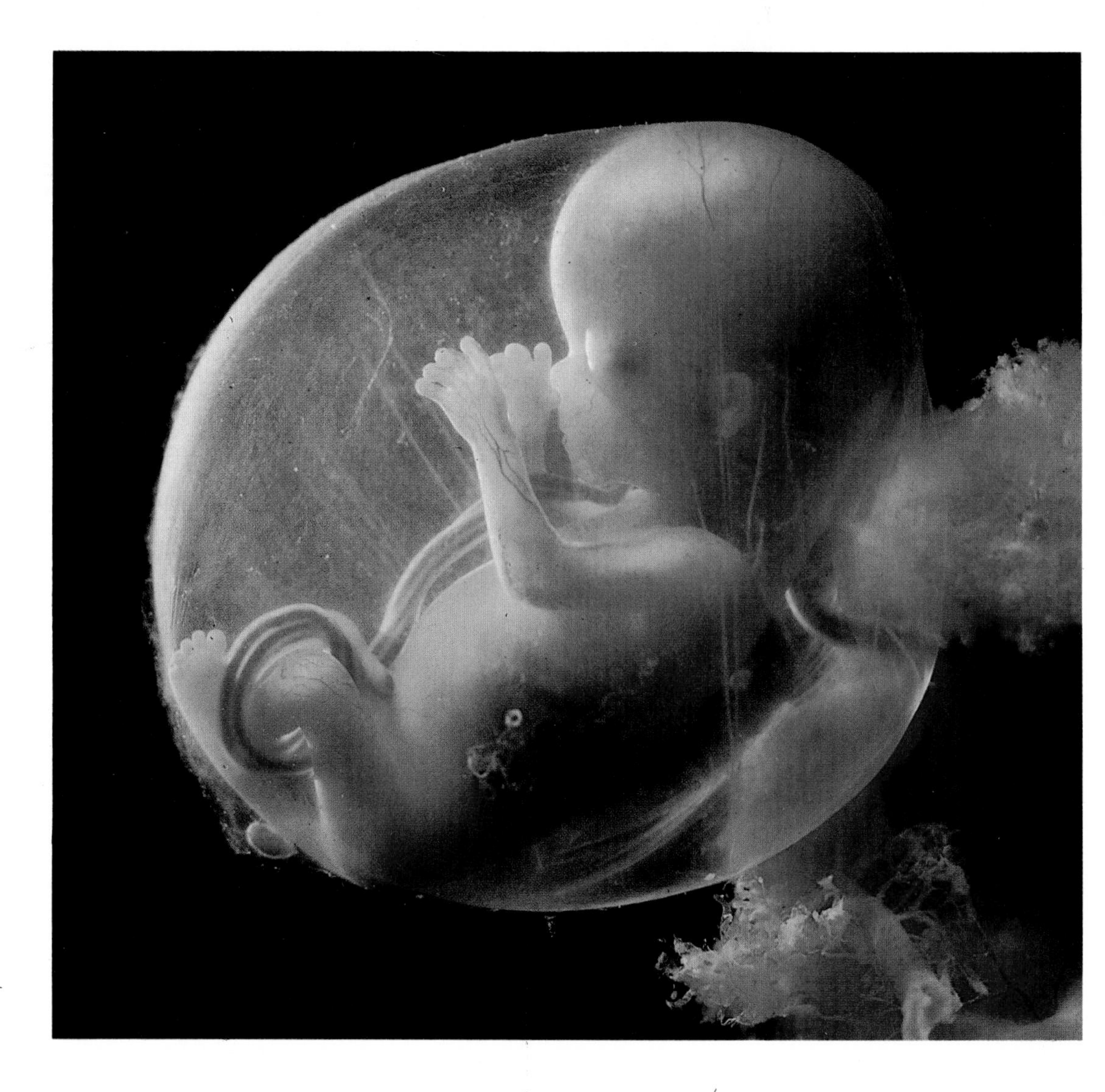

A human fetus, four months old and only six inches long, attached to the placenta, an organ that separates the fetal bloodstream from that of its mother.

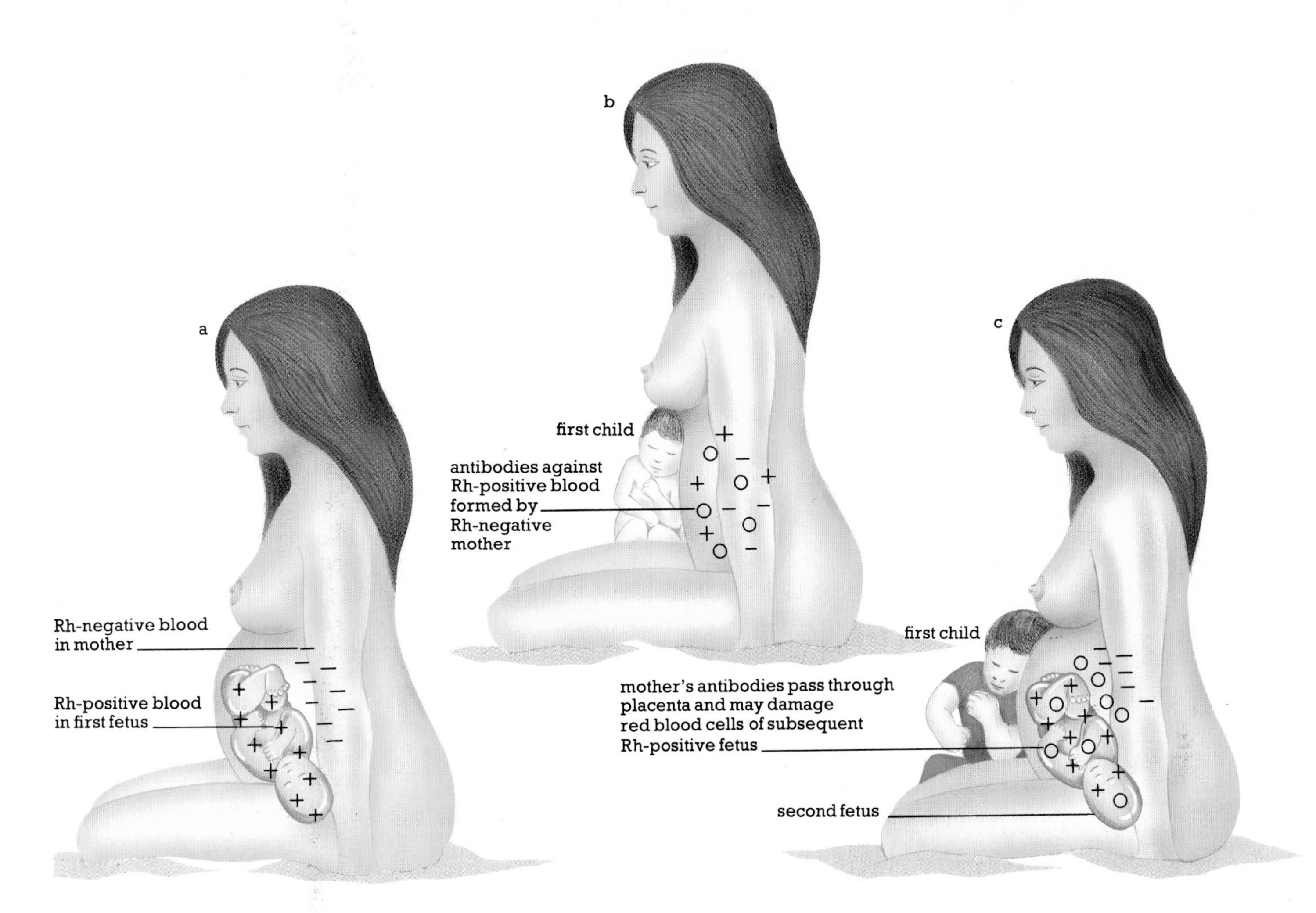

dominant and so an Rh-negative person cannot carry the rhesus factor on either of their chromosomes. If an Rh-negative woman becomes pregnant by an Rh-positive man, the child may be Rh-positive. But this will only cause problems if the Rh factor from the fetus escapes into the mother's bloodstream. This may happen as the result of an accident or at birth. The mother will then produce antibodies against the Rh factor on the fetal red blood cells.

The first child suffers no ill effects. But if the woman has a second Rh-positive child and fetal cells again enter the mother's bloodstream, then she will produce large amounts of anti-Rh antibodies as her immune reaction is increased. These antibodies may then enter the fetal bloodstream and cause clumping of red blood cells, which threatens the survival of the fetus. The baby needs blood transfusions immediately after birth.

These diagrams show how a mother's immune system produces antibodies against the Rh factor, which, like the factors involved in the ABO blood system, is a factor on red blood cells. An Rh-negative mother with an Rh-positive fetus (a) develops antibodies against Rh factor. The first child is born before any ill effects occur (b), but such antibodies may destroy the red blood cells of a second Rh-positive child (c).

Fortunately the genetics of this situation are now well understood and the risk to the second Rh-positive child can be removed. As soon as her first Rh-positive child is born, the mother is injected with a substance that destroys any Rh-positive fetal cells before they can stimulate the production of anti-bodies. Then a subsequent Rh-positive fetus will not be at risk at all.

Sickle-Cell Hemoglobin

In common with nearly all animals and plants, human beings need oxygen in order for their cells to produce vital supplies of energy. Blood circulating through the lungs absorbs oxygen from the air and carries it in the red blood cells to all the tissues and organs of the body. In the blood cells is a protein called hemoglobin, which contains iron and gives blood its red color. Hemoglobin attaches molecules of oxygen to itself and delivers them to the tissues.

Many people in Africa suffer from an inherited disease known as sickle-cell anemia, in which the red blood cells look like tiny sickles. These cells can break easily; they also tend to clump together and clog up the blood vessels, leading to the malfunction of organs such as the brain, heart, and kidney.

The cause of sickle-cell anemia, which frequently leads to death in childhood, is to be found in a minor change in hemoglobin. Hemoglobin is a protein that comprises four chains of amino acids – two called alpha chains and two beta chains. Each alpha and beta chain contains about 150 amino acids and for each type of chain there is a different gene. In cases of sickle-cell anemia the same single amino acid in each of the beta chains is altered. There are just two incorrect amino acids in the 600 or so amino acids in the hemoglobin molecule. The molecules of abnormal hemoglobin tend to link up with each other to produce long rods that can distort the cells into the sickle shape. The altered amino acid is in turn due to a change in just one base in a sequence of about 450 bases in the DNA that constitutes the gene for the beta chain of hemoglobin. The change, though minute, is ultimately responsible for the varied symptoms of the disease. It is inherited.

However, the severe symptoms of sickle-cell anemia appear only in people who inherit an abnormal gene for hemoglobin from both parents. People who have one normal gene and one abnormal

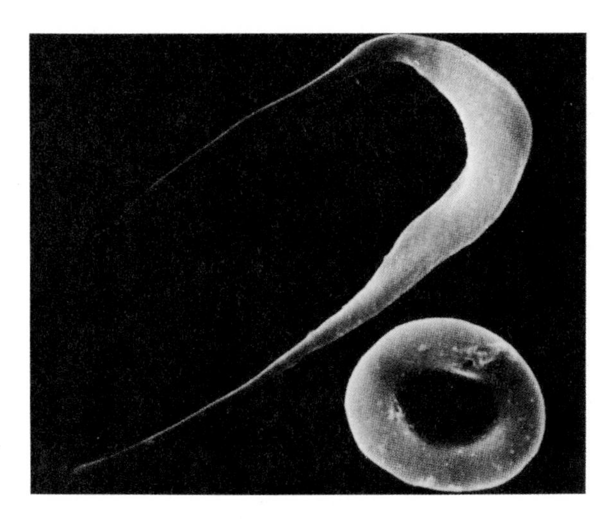

Blood cells from a patient with sickle-cell anemia (left) have the sickle shape from which the disease takes its name. This shape occurs when the level of oxygen in the blood is low and the hemoglobin tends to form fibers. A normal red blood cell (right) is disk-shaped and has shallow indentations on both sides in the center.

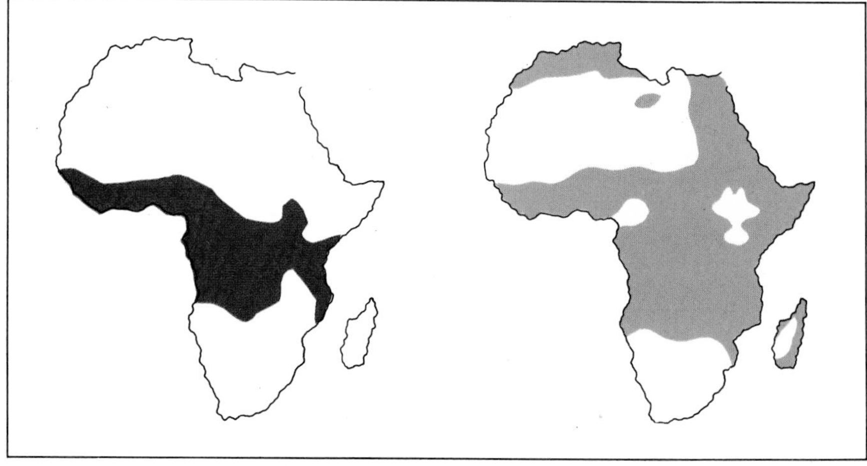

The map of Africa on the left shows the regions of relative frequency in the population for sickle-cell hemoglobin (white indicates lower frequency, brown indicates higher frequency). The map on the right shows the regions where malaria occurs most frequently. As can be seen, the gene for sickle-cell hemoglobin tends to occur in regions where malaria is rife.

gene actually produce both types of hemoglobin. Such people develop sickle cells if oxygen is scarce, for example, at high altitudes. Under such conditions many molecules of sickle-cell hemoglobin stick together, forming rods that pull the cells out of shape. These individuals can pass on the abnormal sickle-cell hemoglobin gene to their children who, if they receive a sickle-cell gene from both parents, are likely to die before reaching adulthood.

In some parts of Africa up to forty per cent of the population have one or two sickle-cell hemoglobin genes. Therefore the chances of a child having two sickle-cell hemoglobin genes, and consequently dying, are quite high. Of course, if a child dies it cannot pass on its genes. However, if a child has only one sickle-cell hemoglobin gene, its chances of survival are very good. Indeed, it has a better chance of survival in these areas than does a normal child. Why should this be?

The answer comes from a rather surprising finding that might appear to be unrelated. Malaria is common throughout much of Africa. This disease is caused by a single-celled parasite that spends part of its life in human red blood cells. It is transmitted from person to person by mosquitoes that feed on human blood. On maps of Africa, the occurrence of sickle-cell anemia and malaria coincide so well as to suggest a connection. This is confirmed by medical research, which shows that people with a sickle-cell hemoglobin gene are relatively resistant to severe attacks of malaria. So in areas where malaria is common, there is some survival value in having a sickle-cell hemoglobin gene. People with one normal and one abnormal gene probably have the best chances of survival, since one gene prevents them from having sickle-cell anemia, while the other protects them from malaria. This explains the high incidence of the sickle-cell hemoglobin gene in the population.

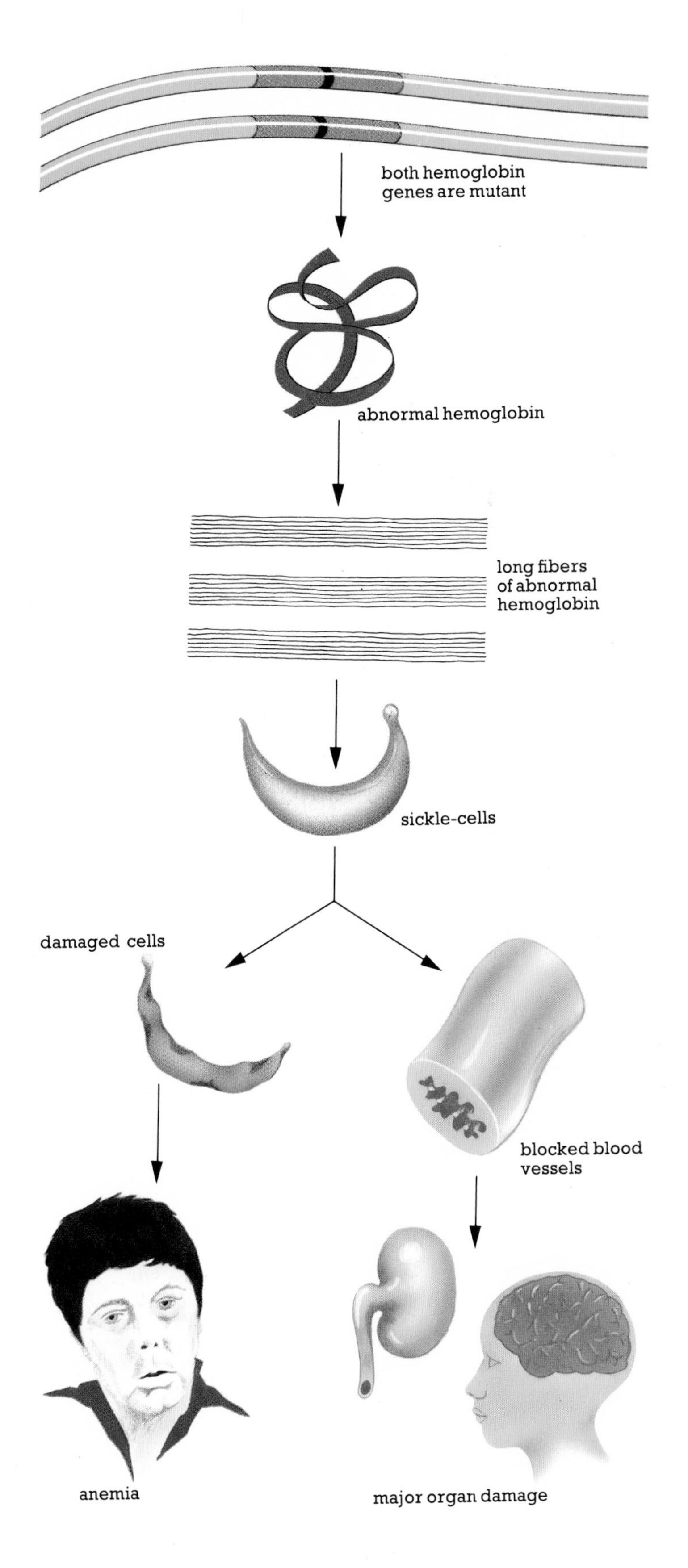

When genes from both parents are in the mutant form leading to sickle-cell hemoglobin (top) they produce an abnormal kind of the protein in which the hemoglobin forms long fibers when the level of oxygen in the blood is low. The result is a chain of symptoms, shown below, that frequently ends in death.

Viruses

Living organisms are normally distinguished from nonliving objects by their ability to reproduce. However, this is not an adequate definition as far as viruses are concerned. In one sense, they are tiny living organisms, which exist and reproduce as parasites inside other living organisms. This relationship often leads to disease in the host organism. A number of human diseases are caused by viruses – smallpox, poliomyelitis, measles and the common cold, to name only a few. Animals and plants also suffer from diseases caused by viruses, and there are even viruses that can infect bacteria. Viruses are extremely small, so small that several thousand arranged in a straight line could fit on a pin-head.

However, viruses are much simpler than other living cells. They are so simple that frequently they can be crystallized just like everyday substances, such as salt or sugar. Indeed, many viruses are made up of just two major chemical components, DNA (or RNA) and protein. How then can these tiny structures, on the borderline between living and nonliving, reproduce?

An interesting answer to this question came from some experiments on a virus that infects plants, producing a speckled appearance on the leaves. In 1933, the American microbiologist Wendell M. Stanley crystallized this virus, known as tobacco mosaic virus, or TMV. It was later shown that the virus consisted of a central core of RNA surrounded by regular blocks of protein. The RNA and the protein blocks could be separated and then put together again in a test tube to produce the active virus capable once again of infecting tobacco plants. Even more remarkably, experimenters found that they could take apart two varieties of TMV virus, distinguished by the nature of their protein blocks, and then rebuild them in new combinations: RNA from variety A with protein blocks from variety B, and RNA from variety B with protein blocks from variety A.

The hybrid viruses were then used separately to infect some tobacco plants, producing crops of new viruses within the plants. The protein blocks in

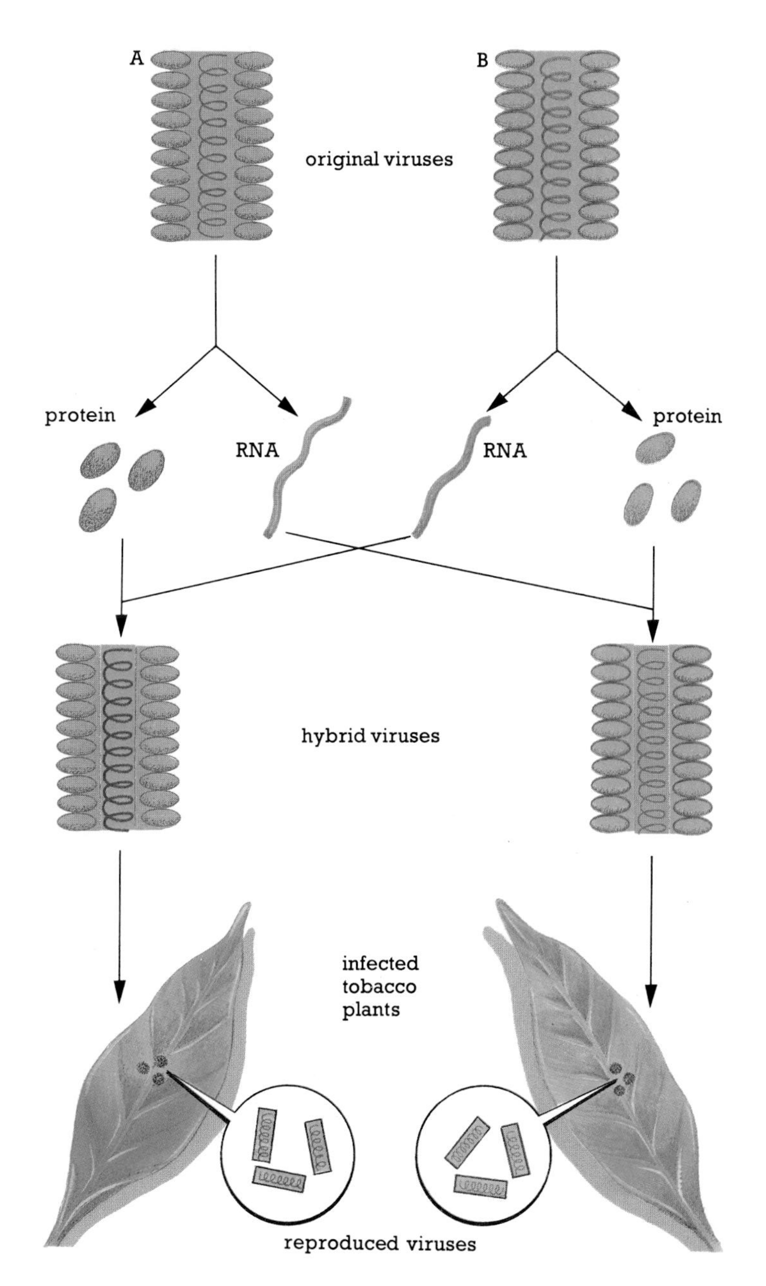

This diagram shows how two varieties of virus, A and B, can each be broken down into their component blocks of RNA and protein, which can then be recombined to form two hybrids. One hybrid acquires the RNA from A (shown in red), with protein blocks from B (shown in blue). The other hybrid acquires RNA from B, with protein blocks from A. If tobacco plants are infected with these hybrids, the resulting viruses, recovered from the leaves, show characteristics that depend on the RNA of the hybrids.

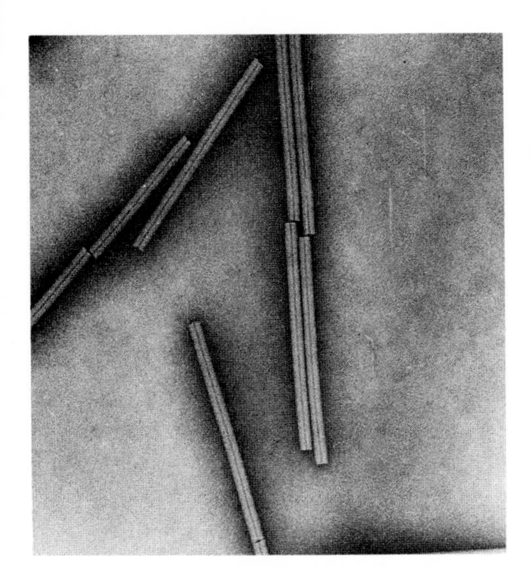
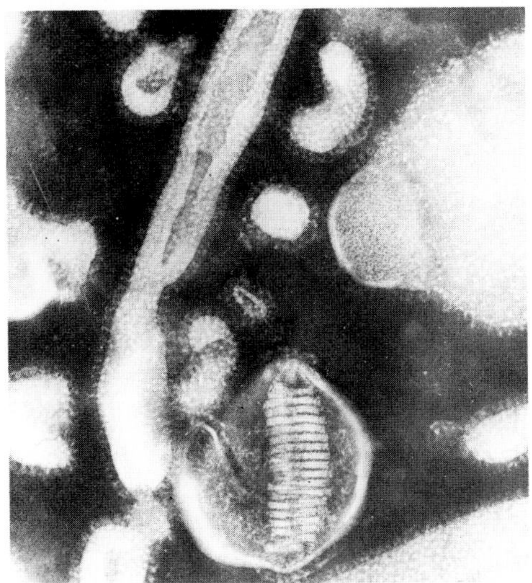

These electron micrographs show the wide differences in shape between viruses: tobacco mosaic virus (left), influenza virus (center) and adenovirus (right). (All viruses are magnified about × 300,000.)

these new viruses always resembled the protein blocks characteristic of the virus from which the RNA originally came. For example, a plant infected with a hybrid virus containing variety A RNA and variety B protein blocks, produced a crop of viruses with variety A protein blocks. It appeared that the RNA contained the instructions for making the protein blocks; in other words, the RNA contained the genetic information.

A virus therefore reproduces inside the plant cell by imposing its own genetic instructions on the cell. It produces more virus genes and the correct, corresponding protein blocks which are assembled together to give a crop of viruses.

Some viruses do not even enter intact into the host cells, but merely sit on the surface of the cell like a space capsule on the moon. They then bore a small hole in the surface and inject their hereditary material. The invaded cell obligingly produces hundreds of new viruses, which break open the cell and go on to infect many more cells – a rapid, economical, and simple method of reproduction.

In the sequence shown below, a virus lands on the outside of a cell (a), bores a small hole and injects its DNA into the cell (b). Once present in the cell, this DNA programs its own reproduction and the making of virus proteins. The proteins and the DNA are then assembled (c) and the resulting viruses break out of the cell (d). Each virus can then go on to infect another cell.

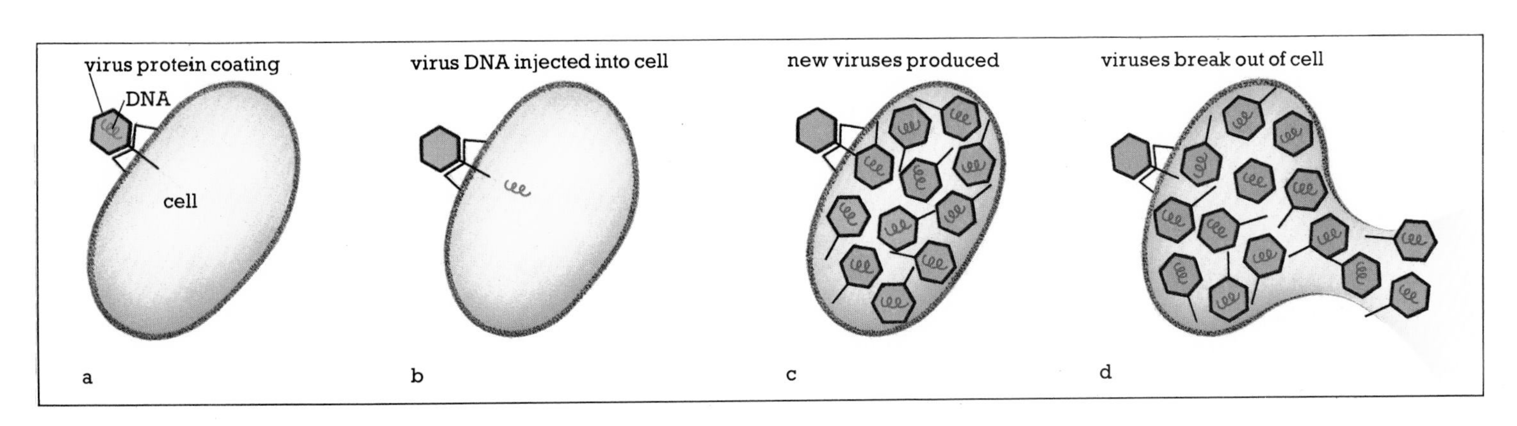

Genes and Bacteria

There are a large number of different species of bacteria, each with its own distinct lifestyle and home. There are bacteria on the surface of our skin, in our intestines, in water, in soil, carried on air currents, on the leaves of plants, even buried deep down in the sulfur beds of Texas. Many of these bacteria are harmless. But some can cause disease, both in plants and in animals: food poisoning, pneumonia, tetanus, and plague (for example, the terrible Black Death, which killed almost half the population of medieval Europe) are all caused by bacteria.

Bacteria are microscopic single-celled organisms that reproduce by simple cell division. Their genes are often present in one long molecule of DNA. This is not wrapped up with histones to form a chromosome, as in animals and plants, but is still reproduced and parceled out, one copy to each daughter cell, during cell division. Bacteria reproduce asexually and are haploid. However, some species of bacteria do have ways of transferring genes to each other – a primitive form of sex. Unlike animals and plants, this bacterial sex does not lead to a fertilized egg and offspring. Instead, some of the bacteria involved simply receive genes from others, which they may then incorporate into their own DNA. If the new genes differ from the existing ones, this leads to genetic change.

There are three basic types of gene transfer in bacteria: transformation, transduction, and conjugation. Transformation occurs when fragments of DNA – a few genes – from a dead bacterium are taken up by other bacteria. In transduction genes are carried from one bacterium to another by viruses. Only in conjugation is there actual physical contact between the bacteria-giving genes and bacteria-receiving genes, by means of a delicate tube called a pilus that is present on the donor bacterium. Because conjugation can take well over an hour, this may break before the process is over, so that only some genes are transferred.

In addition to one long molecule of DNA, some species of bacteria have other much smaller pieces of DNA. Sometimes these, too, can be transferred by conjugation. Unfortunately, these small pieces of DNA often carry genes that resist various antibiotics, for example, penicillin and streptomycin. A single piece of DNA can give resistance to several antibiotics, and so in a single transfer a bacterium can become immune to antibiotics. Even worse, the pieces of DNA can sometimes be transferred from one bacterium to one of another species. This means that the normally harmless inhabitants of our intestines can pass on their resistance to less common bacteria, some of which are far from harmless.

The transfer of resistance to antibiotics is becoming increasingly common. This is probably due in part to

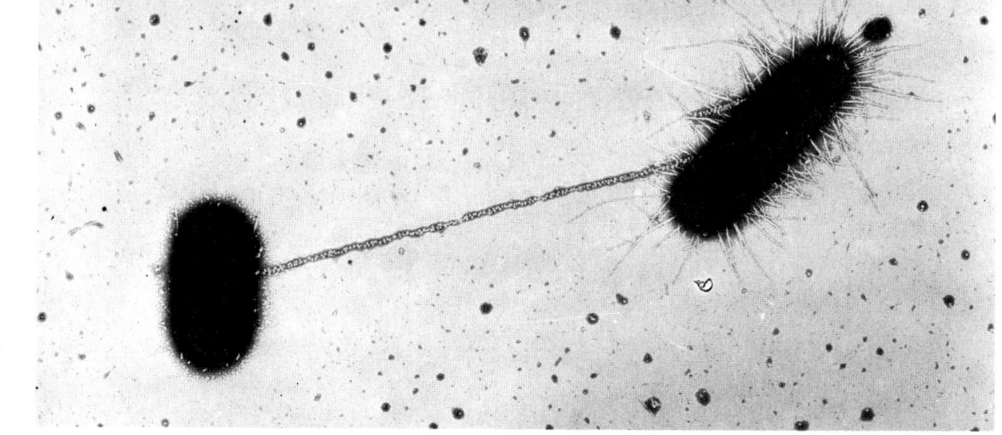

This photograph (magnification × 11,000) shows two E. coli *cells involved in conjugation. One bacterium, the donor (right), passes its genes through a tube, called a pilus, to the recipient bacterium.*

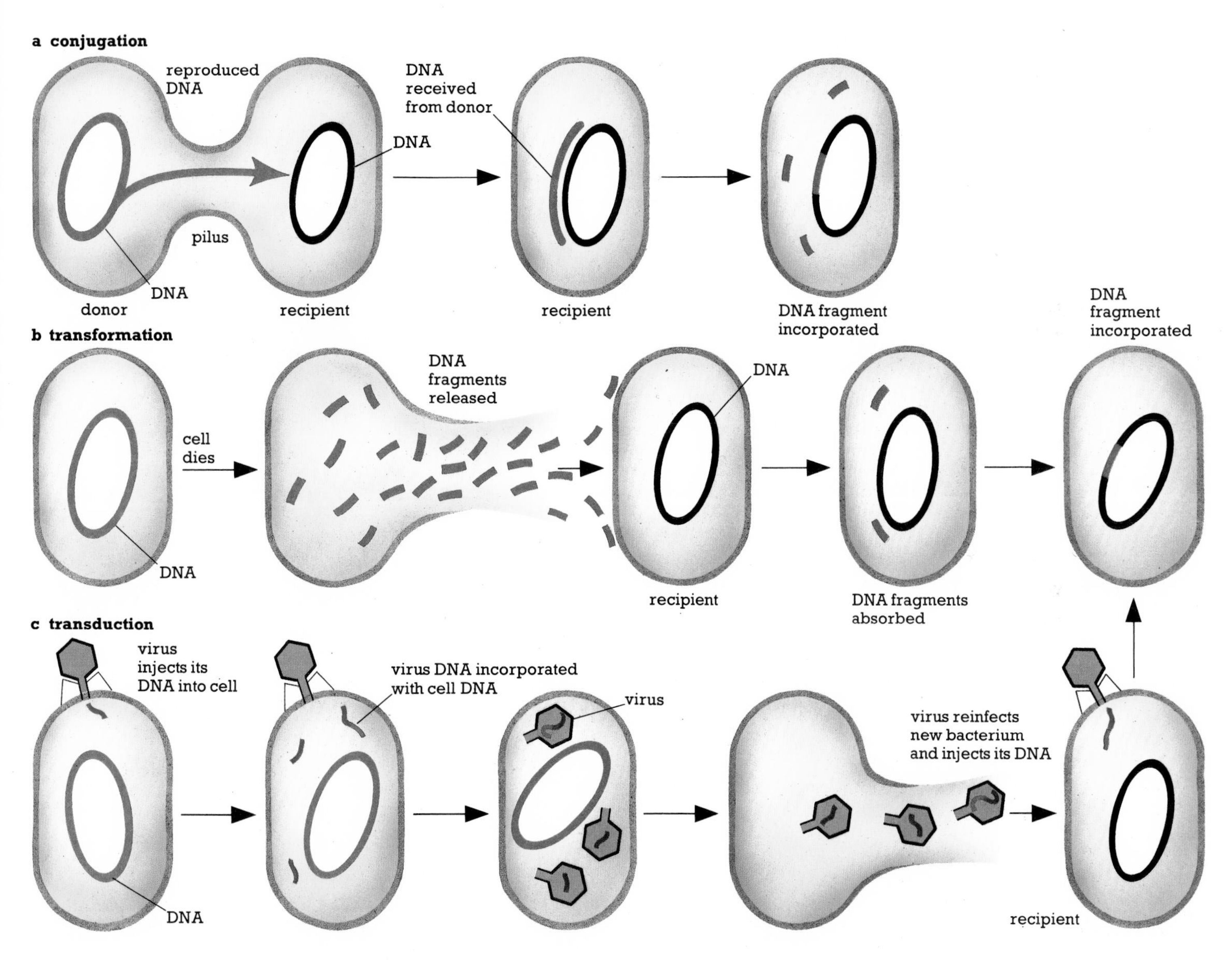

the overuse of antibiotics, both in treating illnesses and in rearing animals for food. The widespread use of antibiotics means that many bacteria, both harmless and otherwise, are killed, leaving the way open for resistant varieties to exploit the available resources, breed quickly, and pass on their resistance. It takes many years of hard work in the laboratory to develop a successful antibiotic, but only a short time for it to become virtually useless if resistance spreads.

Besides conjugation (a), bacteria have two other ways of transferring DNA. In transformation (b), bacteria absorb fragments of DNA released from dead bacteria. In transduction (c), a virus infects a bacterium and picks up some of the bacterial DNA. If the virus then infects another bacterium it may pass on the "stolen" DNA. In each case, if the recipient bacterium reproduces itself, the new bacteria will contain DNA from both donor and recipient.

From Egg to Adult

How does an egg develop into a chicken, or an acorn into an oak tree? Why do apparently similar eggs develop into organisms as diverse as frogs and snails, or turtles and birds?

Some early researchers suggested that eggs or sperm contained tiny preformed organisms, and that following fertilization development was just a matter of growth. Scientists later discovered that the transition from egg to adult is gradual, and that development involves an increase in complexity.

The path from egg to adult form does not always seem to be direct. In some cases the intermediate stages are clearly adapted to the lifestyle of the organism. So, the caterpillar is the stage at which feeding occurs, while the outer covering of the pupa protects the adult butterfly or moth developing within it. In other organisms the purpose of certain developmental stages is not at all obvious. For example, at about four weeks, a human embryo has a

This sequence shows the development of a saturniid moth from egg to caterpillar (top), from caterpillar to pupa (middle) and from pupa to adult (bottom).

A cluster of eggs

A caterpillar hatches out from an egg

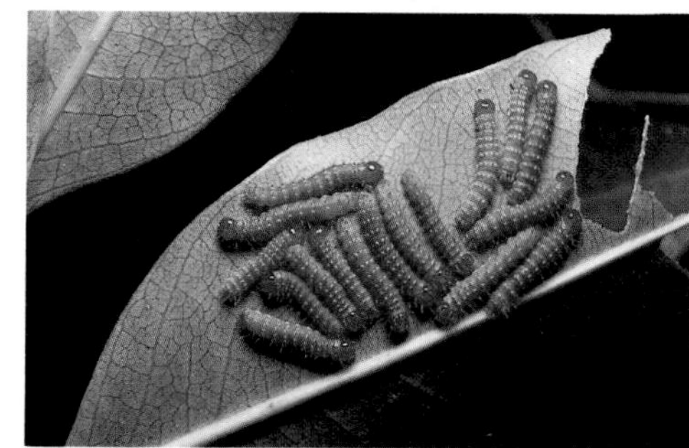

Young caterpillars, or larvae

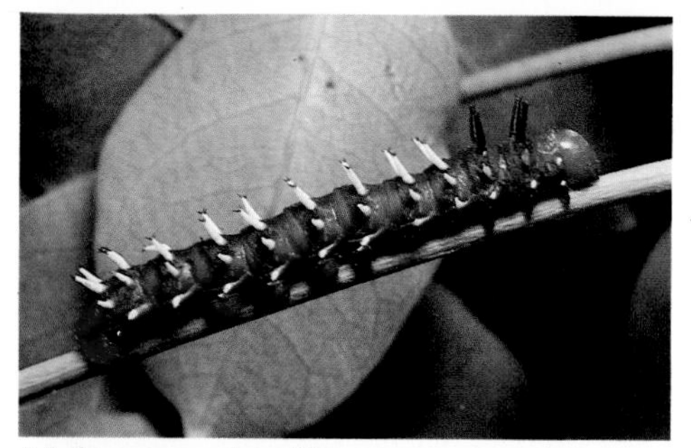

A fully grown caterpillar

The beginning of the pupa stage

A pupa, or chrysalis, after ten months

The pupa splits open

An adult moth shortly after emergence

An adult moth with fully extended wings

tail, which later disappears. This organ has no function, but it may indicate that our ancient ancestors had tails.

Although each species of organism has a unique developmental system, features common to development in most organisms can be picked out. For example, any adult is much bigger and more complex in structure and shape than its egg. The increase in size is accompanied by cell division. In the case of a human, a single-celled egg develops to an adult containing about a hundred trillion cells. Moreover, there are many different types of cells, kidney, liver, etc. These are produced during development by a process called cell differentiation. These different cell types are not just scattered throughout the organism. They are clustered to form distinct patterns as, say, on a butterfly's wings, or shapes like that of a fish's fin. Pattern and shape are formed during development by the process known as morphogenesis.

Development therefore consists of three interacting processes: growth, cell differentiation, and morphogenesis. The precise nature of each process varies from one species to another. Yet in each case the apparently hazardous journey from egg to adult goes remarkably smoothly.

The early development of a human embryo is shown at five weeks (a), six weeks (b), seven weeks (c) and two months (d). At this stage it is no longer called an embryo, but is called a fetus. At seven months (e) the fetus is recognizably human.

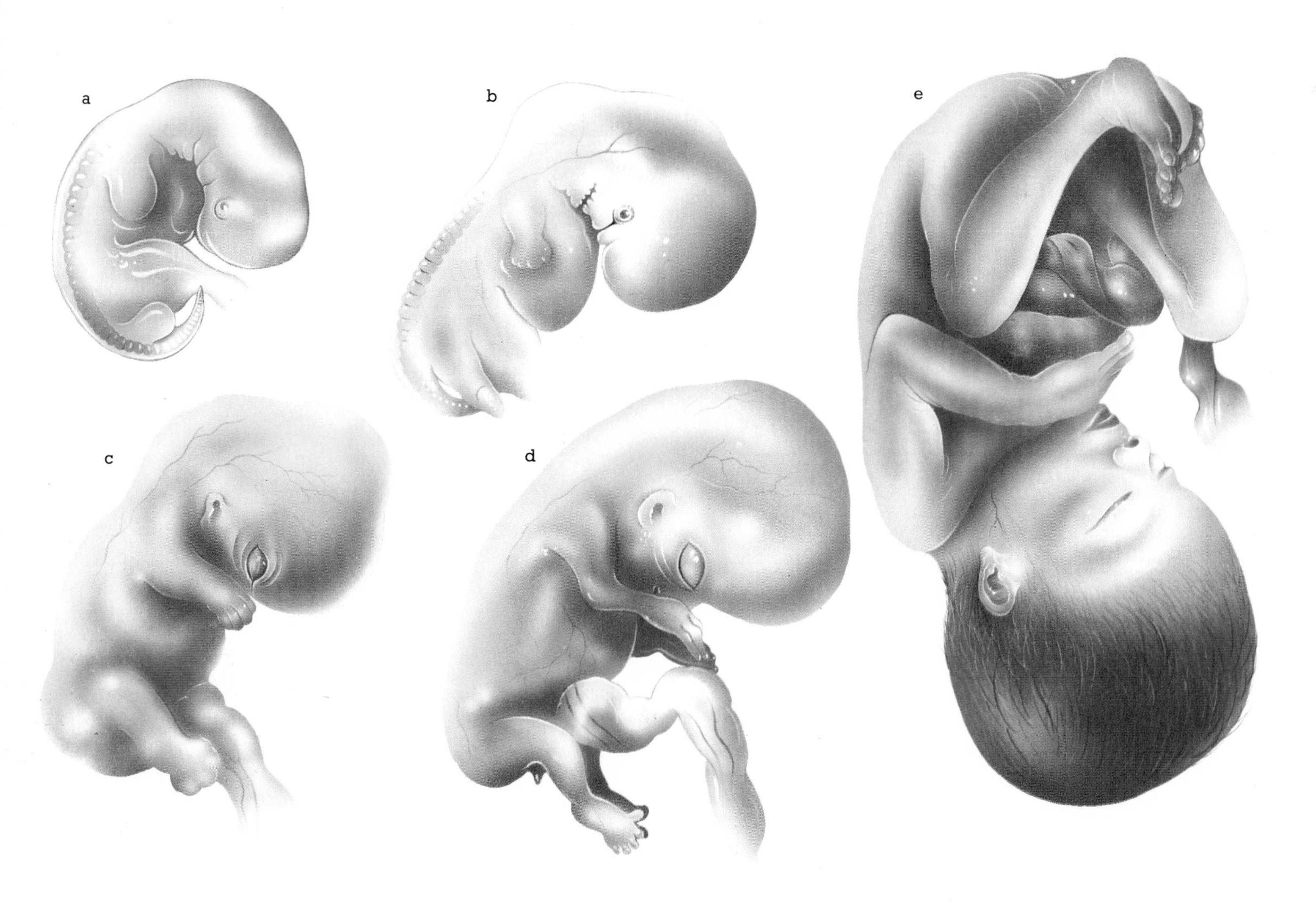

Genes and Development

Both genes and environment play important roles in development. A particular set of genes ensures that a hen's egg always develops into a chick, not a kitten. At a more complex level genes help to determine the precise form of each part of the organism. For example, a mutation can cause human beings to grow extra fingers.

Environment, too, helps to ensure that an organism develops correctly, and an abnormal environment can result in physical deformities. One tragic example involved the tranquilizing drug thalidomide, which modified the environment – the mother's womb – surrounding the developing embryo. Only after several years, and the birth of many thousands of severely handicapped children was it realized that, if taken during the early stages of pregnancy, the drug could cause major physical changes in the development of an embryo. Children with badly deformed arms and legs often resulted.

Even when the environment does not contain such factors as thalidomide, it can still have a drastic effect on embryos. For example, there is a type of mouse that has an extra vertebra in the backbone. This addition apparently depends, not on a mutant gene in the embryo, but on the womb in which the embryo develops. This is confirmed by the fact that transferring a developing embryo to the womb of another variety of mouse leads to offspring who do not have the extra bone.

Environment has another even more interesting role. An organism inherits all the kinds of genes it will ever possess when the mother's egg is fertilized by the father's sperm; yet a fertilized egg is totally different from an adult. The reason for this is that a

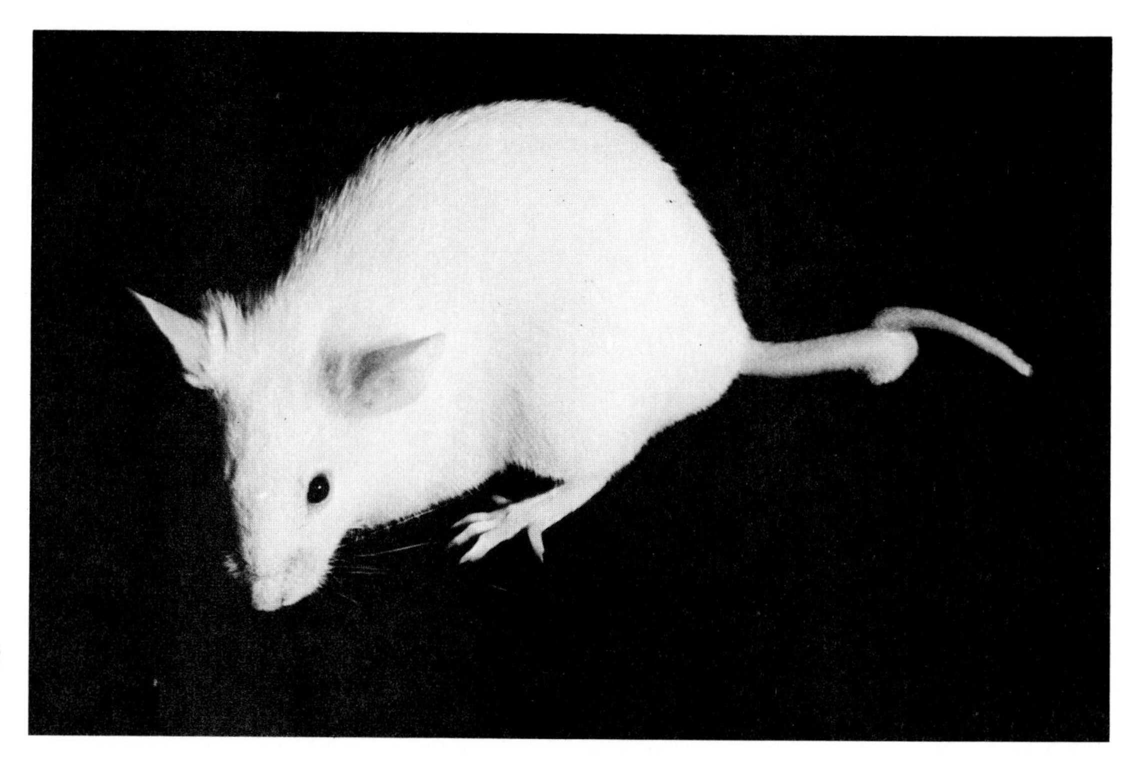

A mutation in this mouse has led to the development of a kinked tail instead of the normal straight tail.

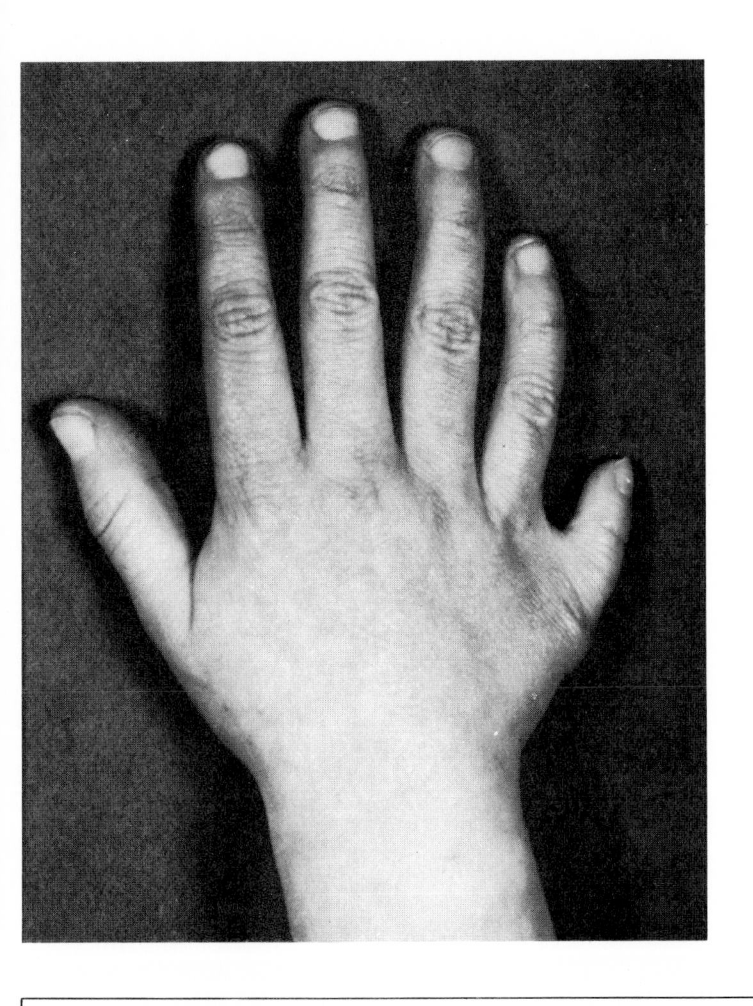

The inheritance of a mutant gene has resulted, at left, in a human hand with one extra finger, a deformity known as polydactyly.

continuous interaction between the environment and the genes leads to an organism's genes being gradually switched on and off as development proceeds. It seems as if neighboring parts of a developing embryo can communicate with each other, so as to keep development along the right lines. For example, the eye and the optic cup, in which the eye sits, are formed from separate tissues during development. Communication between the developing eye and cup helps ensure that they finally fit snugly together.

Such interaction between developing tissues probably depends on chemicals passed between the cells or on physical contact between close neighbors. The messages received in this way by cells probably tell them which particular genes to utilize. Thus, the genes the embryo inherits at fertilization, the different parts of the developing embryo, and the surrounding environment, all interact to produce development.

These chicken embryo wings, viewed through a microscope, show that development can be manipulated. At left is a normal chicken wing, while at right is one that was surgically altered at an early stage. In the embryo a chicken wing forms from a growing stump. If a piece of tissue is cut from the end of this stump and grafted onto the end of a similar stump in another chicken embryo, the wing will develop with parts of the skeleton duplicated.

Genes and Cell Differentiation

In developing from a single-celled egg to a multicellular adult, an organism undergoes many cycles of cell division. The egg divides to give two cells, two become four, and so on in some cases multiplying to billions of cells in the adult. However, the cells of the adult organism are not all the same: different cell types are produced by cell differentiation, which occurs throughout development.

So how can scientists be sure that all cells contain the same genes? The answer comes from a remarkable experiment. If cells from a carrot are placed in a chemical solution, they will grow and divide to give a clump of essentially similar cells, forming what is known as a tissue culture. Researchers found that if they enriched the solution with coconut milk, which is a good source of plant nutrients, growth was particularly vigorous. Isolated cells started to differentiate and eventually a whole new carrot, with leaves and flowers, was produced. A single type of cell from the root thus appeared to have all the genes necessary to produce a whole plant.

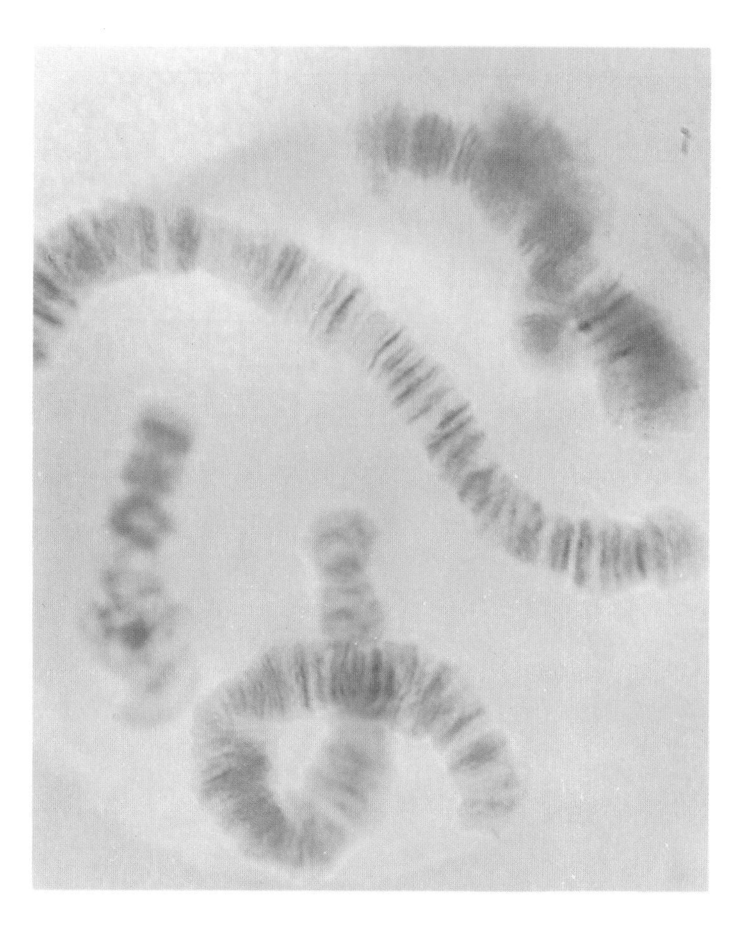

It appears from other experiments performed on frogs that in animals adult somatic (non-sex) cells also all contain the same genes.

If the adult cells of plants and animals contain a full complement of genes, cell differentiation must call for a cell to switch certain genes on and others off, in order to make the particular structural proteins and enzymes that the cell specifically contains.

In most cells the genes that actively produce messenger RNA do so at a time when the chromosomes are invisible. However, in the cells of certain species of flies, the chromosomes are exceptionally thick. Even when the cells are not dividing, the chromosomes can be seen under the microscope as banded rope-like structures with light sections and dark sections. Each light-dark double band probably comprises one gene, and the pattern of bands is a characteristic of a particular species of insect.

Sometimes one or more bands in these giant chromosomes appear rather indistinct, because the tight structure of the chromosome is uncoiled at these points. Called chromosome puffs, these are places where messenger RNA is being manufactured.

In this picture, giant chromosomes from a fruit fly have been magnified and stained. This reveals the characteristic patterns of light and dark crossbands along their length. Each light-dark double band may be a single gene.

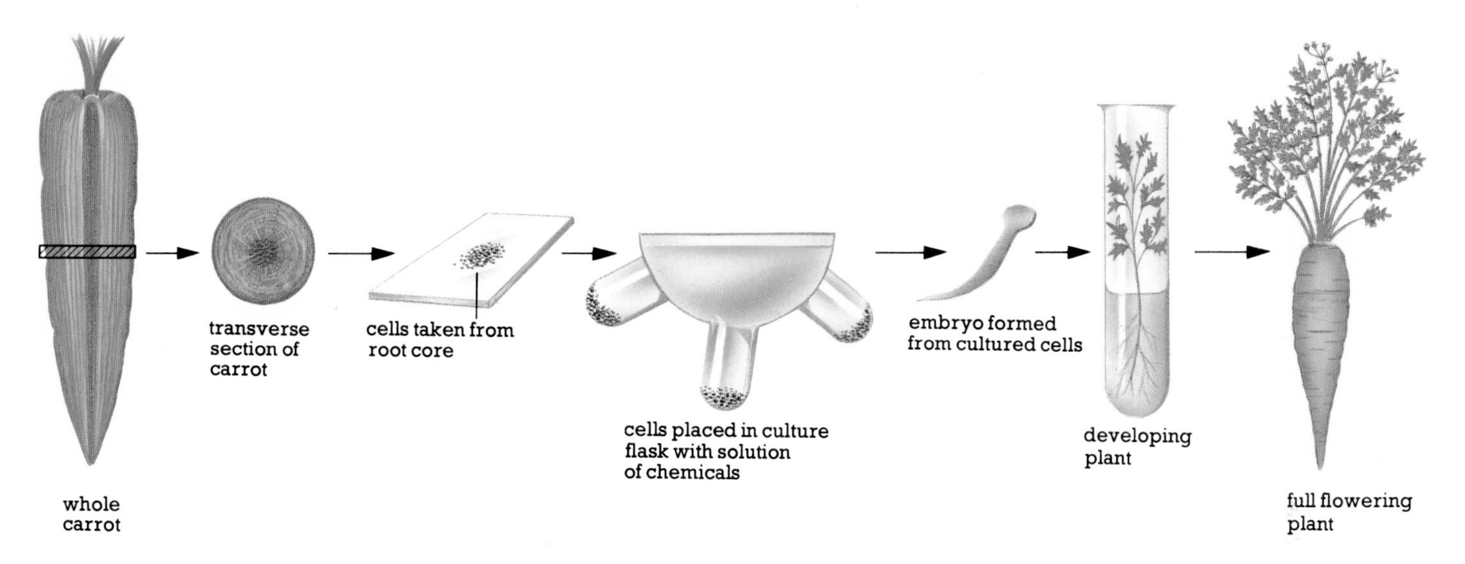

If placed in a suitable solution of chemicals, a single cell taken from a carrot can eventually give rise to a whole carrot plant.

Interestingly, the position of these puffs alters during the life of the cell. Moreover, in different types of cell, puffs are at different places, that is, on different genes. These puffs therefore suggest that, as a cell develops, some genes in it are switched on while others are not. It also suggests that different genes are active in different types of cell.

No one is certain what controls gene action during cell differentiation, but there are a number of possibilities. For example, geneticists believe that developing cells in animals and plants may receive chemical messages from their surroundings, perhaps from other cells. Such messages might interact with molecules that lie on the DNA and thus help to switch genes on and off. Switching might also involve some unfolding of the chromosome as appears to occur during puffing. We must await the results of further research work in the laboratory to find the answers.

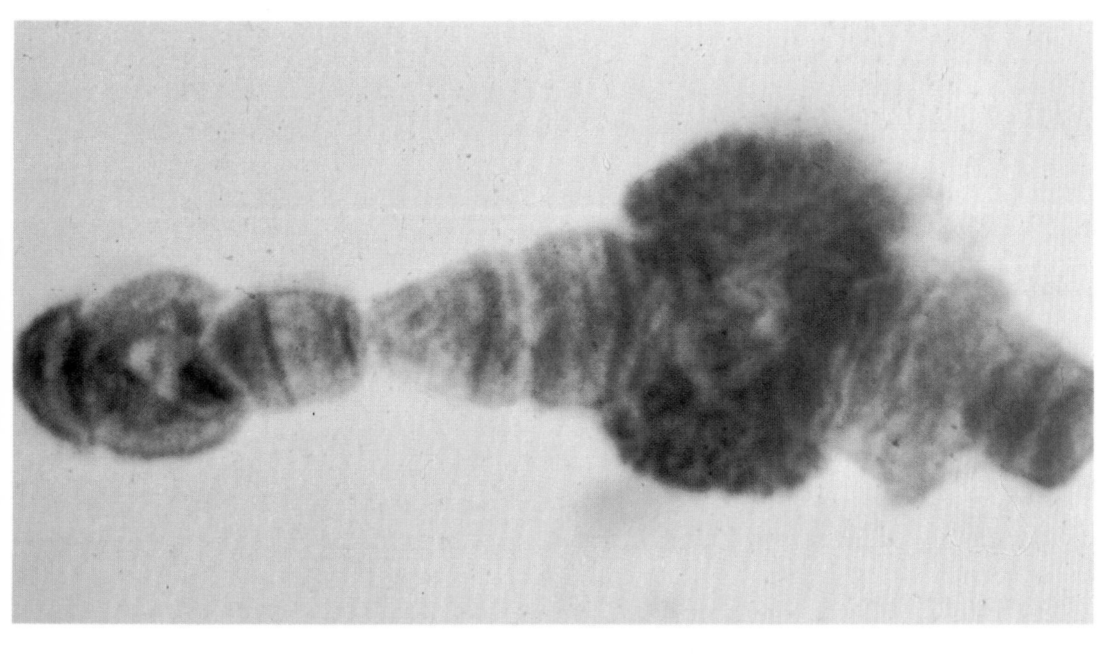

This so-called giant chromosome, magnified many times, is from a salivary gland of a midge. It has been colored by a dye that stains DNA pink and RNA purple, which confirms that RNA is made exclusively in the chromosome puffs. These puffs appear at different points at different times, indicating that genes can be switched on to produce messenger RNA, then switched off again.

Morphogenesis

A list of the different types of cells in a hand – such as skin, blood cells, cartilage, bone, nerve cells – is the same for a foot. The differences between hands and feet then cannot be fully explained just by understanding what genes may be switched on and off during cell differentiation. These differences occur through an amazing, chemically influenced process called morphogenesis, that controls the way in which cells become organized into various patterns and shapes. A very important part of morphogenesis is the fact that cell differentiation and growth take place at the same time, creating patterns of cells that start as parts of an embryo and become organs and tissues as different as brains and hearts, the iris of the eye or the skin of a hand.

To visualize the process of morphogenesis, think of a single sheet of identical cells, and imagine that a chemical is poured across this sheet of cells from left to right, seeping into the cells as it flows across them. The amount of chemical that remains in the cell will decrease from left to right just as paint from a paintbrush gets thinner in passing across a surface. At the same time cells are receiving varying amounts of the chemical, or chemicals, they are growing at different rates. And if, as many scientists believe, the level of chemicals in the environment of the growing organism determines the switching on and off of certain genes in the cells, the cells will change in visible patterns from left to right. No matter how complex the pattern in which the chemical action occurs – such as in a growing hand – gene switching and consequent cell changes will conform.

Morphogenesis also occurs through cell movement, which may be quite dramatic at certain early stages of development. For example, cells move from one region of an embryo to take up positions in other regions, and this can lead to new shapes and patterns. The cause of this movement is not fully understood. In some cases it may be in response to a chemical signal that some cells receive from other cells, essentially saying, "Come here" or "Go away." Or cells may merely wander around and eventually stop when they come into contact with compatible

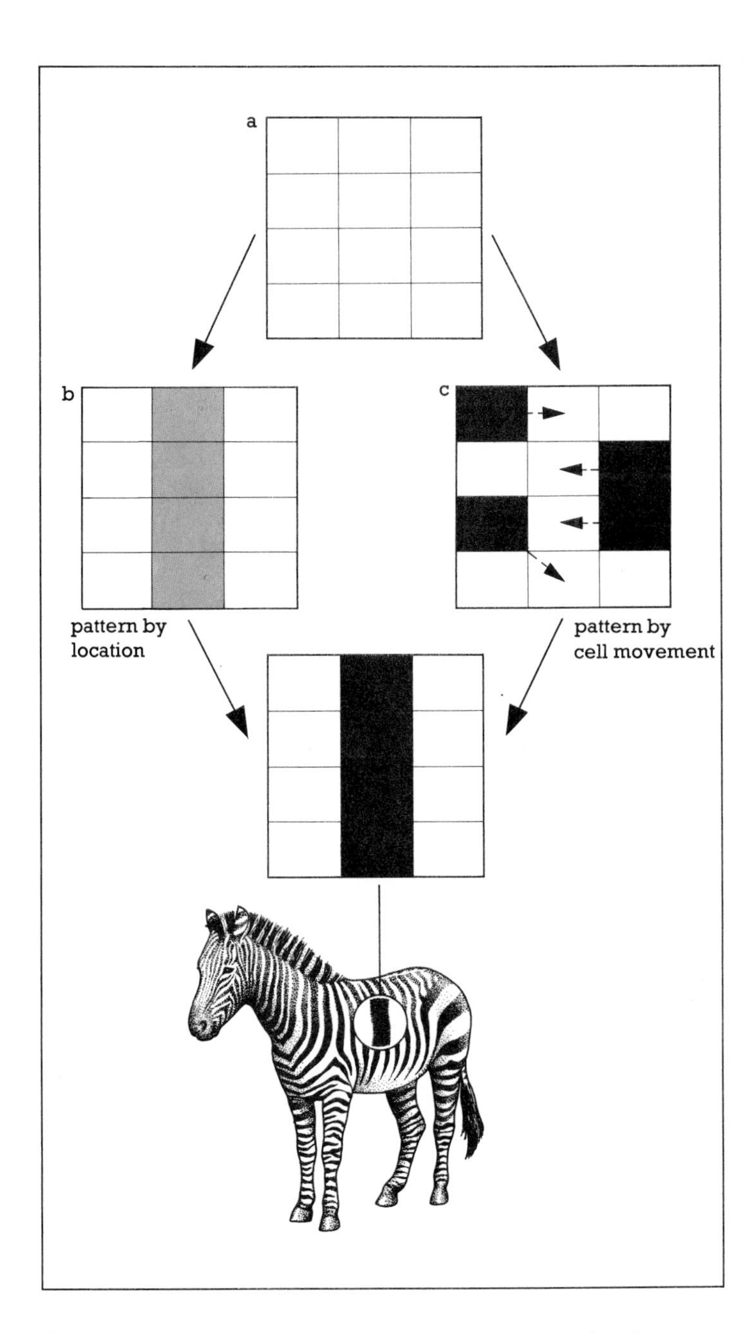

A schematic pathway leading to a pattern of stripes, such as those on a zebra, could be formed in one of two ways. Individual cells (a) produce pigment according to their location (b). Alternatively, the cells become committed to form pigment and then move into position (c).

The intricate patterns of this Monarch butterfly are the result of morphogenesis and cell differentiation during development.

As a stump grows into a foot in a developing embryo, cell death, which occurs in the regions colored red, helps shape the toes. Such cell death is genetically programmed. In a chicken, cell death completely separates the toes, while in a duck some cells, rather than dying, form a web between the toes.

neighboring cells, perhaps those with a special surface like their own. In this way, similar types of cells may cling together to form tissues and organs. Both cell movement and the nature of the cell surface depend to some extent on the types of molecules present on the cell surface, and this ultimately depends on the influence of certain genes. So by modifying the cell surface, genes may control the movement of cells, and their eventual destination.

Along with cell growth, cell death is also an important element in morphogenesis, though in the case of growth, the process is easier to understand. For example, if growth occurs more rapidly in one region of an embryo than in another, it can change the shape of the embryo. Death may not play an obviously active part in development, but the death of specific cells can be vital for shaping certain organs.

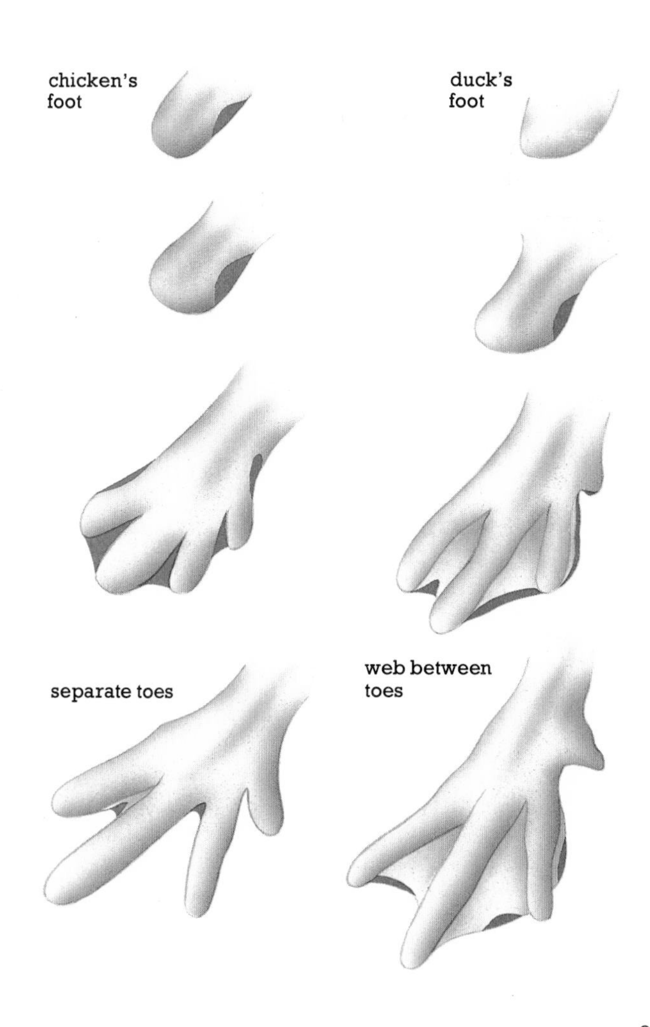

Cellular Slime Molds

When engineers are designing and building a complicated structure, such as a bridge, a ship, or an aircraft, they often build a model. As far as possible, the model is an exact miniature of the intended structure. The model enables engineers to test the design for such things as strength and stability. If necessary, the model can be modified and the design improved before the actual structure is built.

In biology, too, scientists use models, although they do not build these models themselves. Instead they use relatively simple biological systems that resemble more complex ones. Such model systems provide scientists with valuable clues and indicate useful questions to ask about how complex systems function. What is learned from model systems can be applied to the investigation of complex ones.

Development is one of the complex phenomena that can be studied by the use of models. Since in many organisms growth, cell differentiation, and morphogenesis are all going on at the same time, the large variety of cell types, tissue shapes, and patterns makes it hard to sort out what is happening.

One group of simple yet bizarre organisms has proved particularly useful in studying many features of development found in more complex organisms. These are called cellular slime molds. For much of its life, a cellular slime mold is a single-celled organism. It is an amoeba that lives in the soil and feeds on bacteria by engulfing and digesting them.

An amoeba grows and reproduces by simple cell division, but sooner or later it uses up all its food supply. At this point, it dramatically changes its way of life. It joins with hundreds or thousands of other hungry amoebas to form one of a series of streams that converge on a central point. Within a few hours the streaming amoebas collect to form a mound of cells – a small multicellular organism.

This mound then passes through a remarkable sequence of stages. The first, called a slug because of its superficial resemblance to a garden slug, has a head and a tail end, and can detect and move toward light. After a few hours, the slug changes shape and a tiny fruiting body, or reproduction center, is formed. This consists of a mass of cells, called spores, perched on top of a fine stalk about one-tenth of an inch high, composed of cells that, like those of a plant, have rigid cell walls. When fresh supplies of bacteria are available, the spores germinate and amoebas emerge, in much the same way that a plant shoot breaks out from a seed. The cycle of feeding, growth, and division then begins all over again.

The photographs below show three stages in a cellular slime mold's development: an early mound of cells (left), a slug (center), and a mature fruiting body on top of a stalk (right).

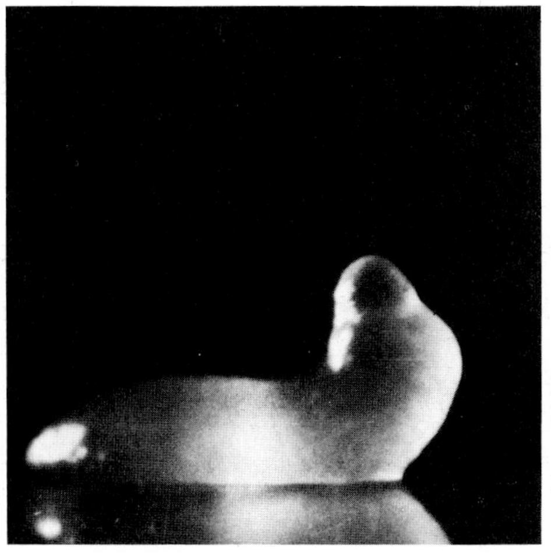

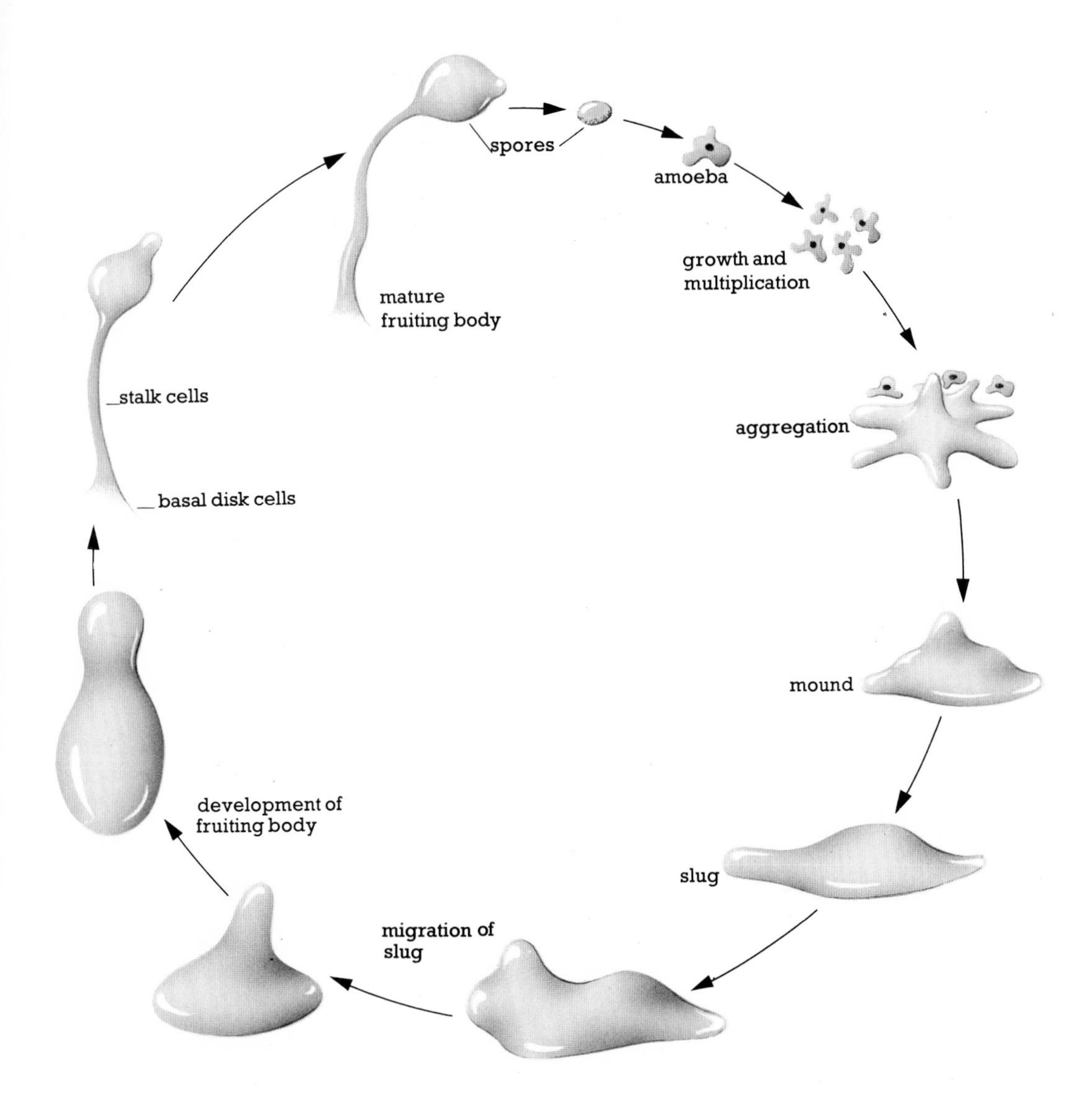

The life cycle of a cellular slime mold (left) begins in the soil with a single-celled amoeba, which grows and multiplies. When food is depleted, individual amoebas aggregate to form a mound of cells. This mound goes through a series of changes in state, including the slug stage, and culminates in the formation of a fruiting body, which comprises a ball of spores on top of a stalk.

This simple life cycle demonstrates the three principal processes of development found in more complex organisms. The first, growth, is restricted to the individual amoeba; during the multicellular stages the cells do not grow and divide. The second, cell differentiation, happens when cells that were initially just amoebas change into either stalk cells or spore cells. Finally, morphogenesis occurs in the transformation from a simple mound of cells to the final fruiting body.

What causes the original solitary amoebas to stream together? How do they know where to go? One factor is a chemical that each cell receives from other cells. The cells come together by following the source of this chemical. But why do they stay together in a multicellular mound? The explanation is probably that changes have occurred on the surface of each cell, so that the cells become sticky and consequently cling to each other.

The coming together of the amoebas to form a multicellular organism depends to some extent on genes; indeed, mutants occur that fail to make a stable mound. In yet other mutants, a mound is formed but development proceeds no further. By unraveling the genetics underlying these mutants, scientists might well learn more about what controls development, not only in slime molds but, ultimately, in higher organisms as well.

Chicken or Egg?

How does development start? Is there anything in the egg, perhaps even before fertilization, that contributes to the various changes that occur later on?

A possible answer to these questions comes from the study of sea urchins, common animals found near rocky shores in many parts of the world. Sea urchins are able to produce large numbers of eggs, which can be fertilized with sea urchin sperm in the laboratory. Once fertilized, the eggs develop into larvae, and eventually into young adults.

Sea urchin eggs can be cut in half with a fine glass needle. Instead of dying, these half eggs become rounded. They can then be fertilized and left to develop. The character of the development depends upon the direction of the cut. If a cut is made vertically, both halves of the egg can produce a normal larva. But if a horizontal cut is made, only the lower half can develop.

This experiment suggests that, even in an unfertilized egg, different regions contain different substances capable of affecting later development. These substances would not be genes, but would interact with genes as the egg develops and divides into further cells. Although each new cell receives a similar set of genes, each could find itself in a different type of cytoplasm, in different regions of what was cytoplasm of a single, unfertilized egg cell. We can speculate that these cytoplasms differ in chemical substances, capable of determining whether particular genes are switched on or off. So depending on which cytoplasm surrounds them, the cells will begin to differ from each other; that is, cell differentiation begins.

In the sea urchin egg the boundary between different types of cytoplasm runs across the egg. In normal development, the first round of cell division,

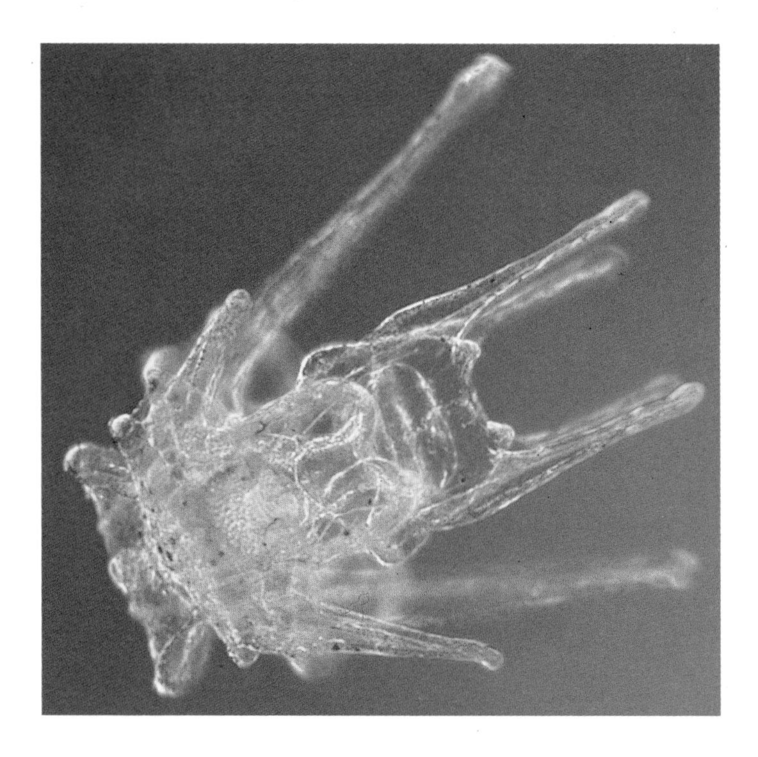

From a single cell, a sea urchin develops into a larva (above), then into an entirely different looking adult (above). The study of sea urchins has provided many answers about development in eggs.

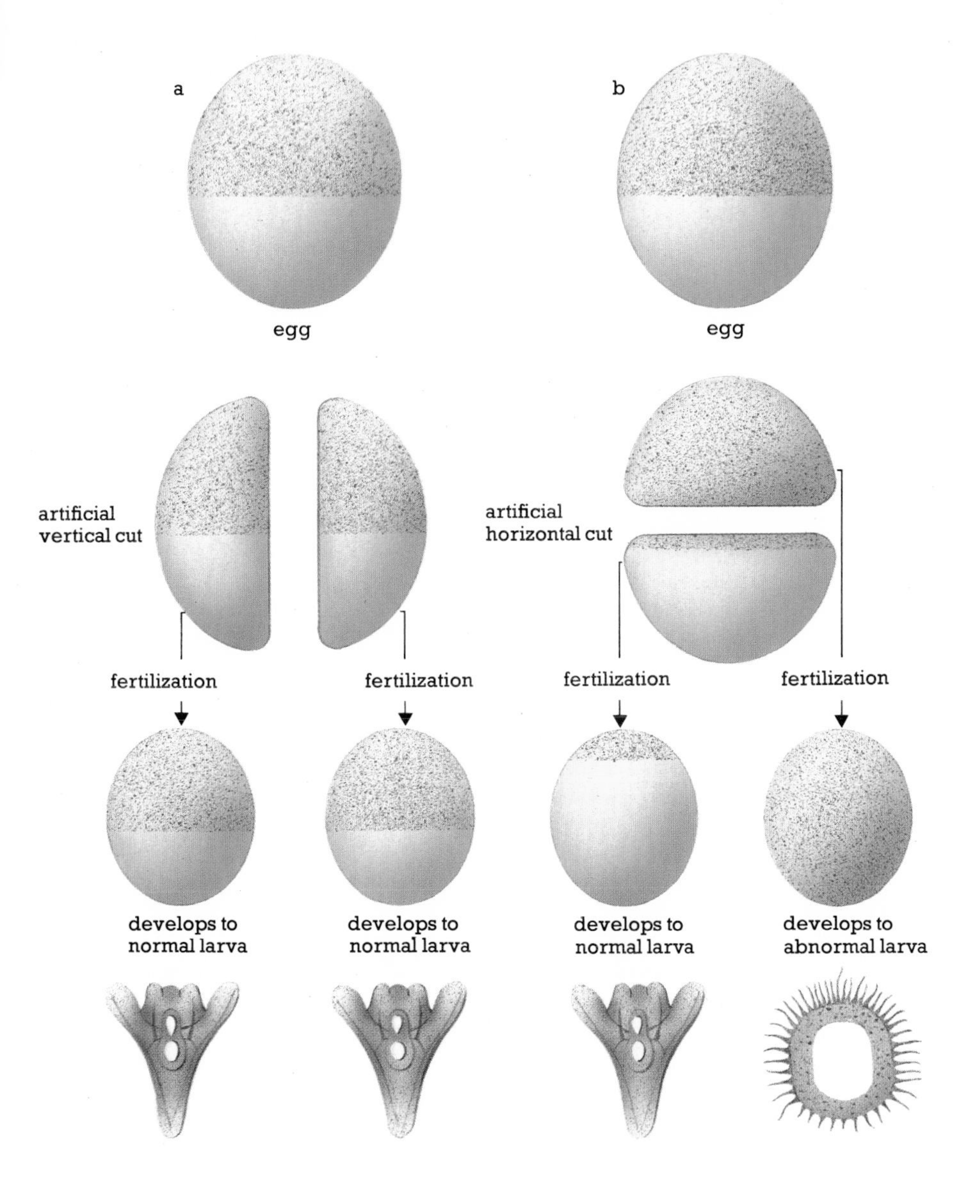

If a sea urchin egg is artificially cut in half vertically (a) and then both halves are fertilized, each can develop into a normal larva, just as an entire egg does. However, if a sea urchin egg is artificially cut in half horizontally (b) and then both halves are fertilized, only the lower half can develop to a normal larva. The upper half forms a very abnormal and incomplete larva. These results strongly suggest that the upper and lower parts of this egg are already different from each other before fertilization and that both parts contain substances necessary for complete, normal development.

giving two cells, runs between the top and bottom. Each one of these cells contains a part of both the top and bottom of the egg. This means that if we separate these first two cells, and allow them to develop, each one will grow into a complete larva. Sea urchins are not unique in this respect. In many organisms each cell in an early embryo can give rise to a whole organism. Indeed, human identical twins might arise from the separation of cells in a two-celled embryo. But in all cases, cells of a developing embryo eventually begin to differentiate, so that an isolated cell can no longer produce a whole organism.

If the beginning of cell differentiation depends on structures or substances in the egg, we need to know how they got there. The answer is probably that they come from the mother even before the egg itself has been fully formed. For example, in some organisms the position of the egg in the mother's ovary determines which end of the egg will give rise to the head end of the adult that eventually forms from it. In the fruit fly, a mother having a mutant gene can produce an abnormal egg, and irrespective of the genes brought in by the sperm, such an egg develops to give an embryo with two tail ends and no head! This suggests that, in normal eggs, some property of the egg influences the genes, which in turn influence development. So the development of a chicken depends partly on the structure of the egg it comes from; and the structure of the egg depends on the chicken it comes from.

Somatic Mutations

When geneticists study mutation they are usually interested in changes to genes in the sex cells, that is, eggs and sperm, or pollen. These changes can be passed on at fertilization and may appear in generation after generation. But mutations are not confined to sex cells. They can also occur in body, or somatic, cells of the organism. Because such mutations are not in the sex cells and are thus not passed on at fertilization, they generally survive only for the life of that one organism. They can, however, have interesting effects. A somatic mutation that occurs in one cell will be passed on by cell division to descendants of that cell. If this occurs early in an organism's development, the mutant cell will produce a large number of descendants, and the effect of mutation may be considerable.

A somatic mutation can lead to areas of odd-looking cells, for example, white feathers on a normally black crow, or white sectors in the otherwise red eye of a fruit fly. People sometimes have eyes of different

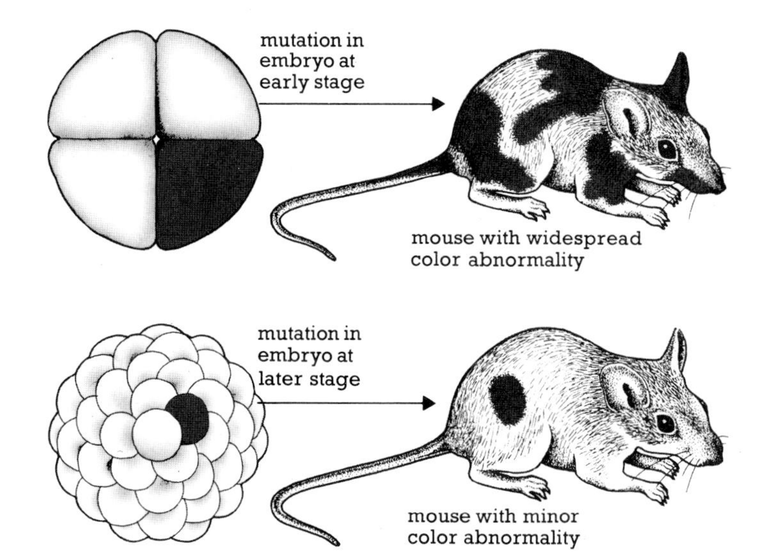

The effect of a somatic mutation depends on when it occurs during cell differentiation. When it takes place early in the development of the mouse (top) its effect is widespread, but when it occurs later in development the effect is limited.

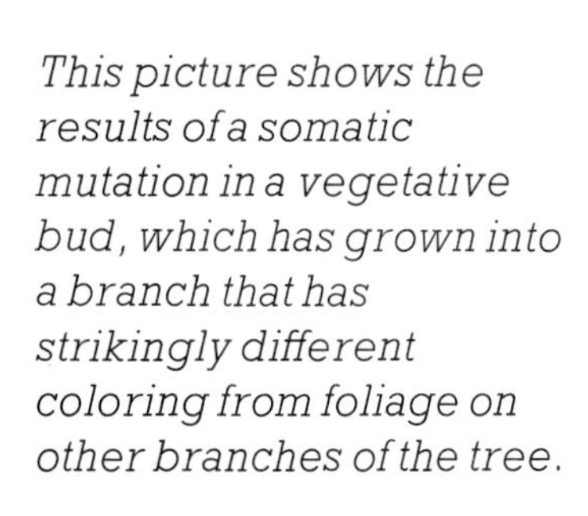

This picture shows the results of a somatic mutation in a vegetative bud, which has grown into a branch that has strikingly different coloring from foliage on other branches of the tree.

colors, which may be caused by somatic mutation in a cell that was the ancestor of one iris. A similar mutation occurring in a later stage of development might cause just a small sector of an iris to be differently colored.

Even though somatic mutations do not pass into sex cells, they can sometimes be passed on to future organisms through human intervention. One such case is when plant breeders take cuttings or graftings to develop new varieties of fruit that have arisen by somatic mutation. Somatic mutations, or what gardeners call bud sports, can occasionally give rise to particularly large or tasty fruits, for example, the Delicious apple and the Washington navel orange. On first occurrence, one such fruit – or bud or branch – is only a small part of the whole tree, but it can be propagated by taking cuttings and grafting them to produce trees with the new characteristic.

Some scientists have suggested that somatic mutation may be a possible cause of cancer. The theory is that a mutation occurs in a somatic cell so that it, and its descendants, divide rapidly and uncontrollably. However, this suggestion is very debatable. Cancer is a complex disease and probably has many causes. Somatic mutation might possibly be responsible for some forms of cancer. But because any genetic change might be very small and hard to detect, it will probably be a long time before the hypothesis can be clearly demonstrated or disproved.

The sharp division between light and dark petals in this chrysanthemum is quite a common form of somatic mutation.

Somatic mutation at a relatively early stage of development is probably responsible for the different colors of this cat's eyes.

The Theory of Evolution

One of the more amazing aspects of the world of living organisms is its huge variety – literally millions of different species. But some species resemble each other more than others; indeed, some seem to be very closely related. A raven is more similar to an ostrich than it is to an elephant, but it is even more similar to a crow. In fact, it is possible to draw up a kind of family tree that relates organisms on the basis of their similarities and differences. A whole area of biology called taxonomy is concerned with specifying relationships between different species.

All living organisms have much in common at the microscopic level of cells and their component structures. But how did present-day organisms arise, and why are they apparently related but in differing degrees? The theory of evolution suggests that all living organisms, including human beings, are the result of billions of years of change.

The theory helps to explain the underlying cellular relationships between all present-day species by suggesting that they have all evolved from a single ancestor, probably itself a primitive single-celled creature. It makes sense of the similarity between structures in quite distantly related organisms – for example, a bird's wing, the human hand, and a whale's flipper. Evolution also explains such apparently useless organs as our appendix or the lobes of our ears as leftovers of history, that may have had a specific function for our distant ancestors. But what caused evolution, and why do certain organisms now populate the earth?

The most widely accepted answer is that proposed by the British scientist Charles Darwin in his famous book, *On the Origin of Species by Means of Natural Selection*, published in 1859. Darwin suggested that, at any one time, each organism is competing with others – including members of its own species – for such things as food, shelter, and members of the opposite sex with which to mate. Success means survival and the production of offspring. Within a species all organisms are basically similar, but minor differences, some of which are inherited, exist

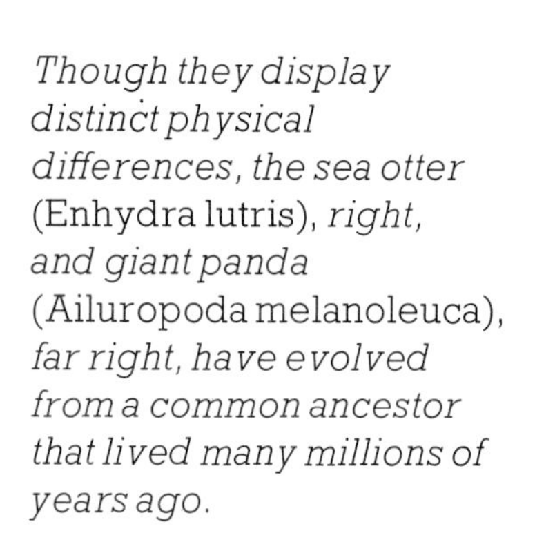

Though they display distinct physical differences, the sea otter (Enhydra lutris), *right, and giant panda* (Ailuropoda melanoleuca), *far right, have evolved from a common ancestor that lived many millions of years ago.*

between individuals. In many cases the differences may be harmful to the organism, making it less well equipped to fit its environment and reducing its ability to compete with members of the same species.

On the other hand, an inheritable change may occur that gives an organism a better chance of surviving. This individual will pass on its advantageous characteristic to its offspring, which will stand a higher chance than normal of surviving, breeding, and producing further, altered, descendants. For example, a chance mutation resulting in a better-camouflaged insect would improve its chance of survival and hence its chance of producing offspring, that would also have the improved camouflage. In this way the descendants of the altered organisms may eventually outnumber the unaltered organisms.

The initial change in the individual is just a matter of chance. Whether that change gives an organism a better chance of survival depends largely on the natural environment. Darwin therefore called this process "natural selection." A change in environment can favor one type of organism rather than another, and an organism that fits the environment better is more likely to survive. Some of the lost organisms of evolution, such as the dinosaur, may have succumbed to sharp changes in climate.

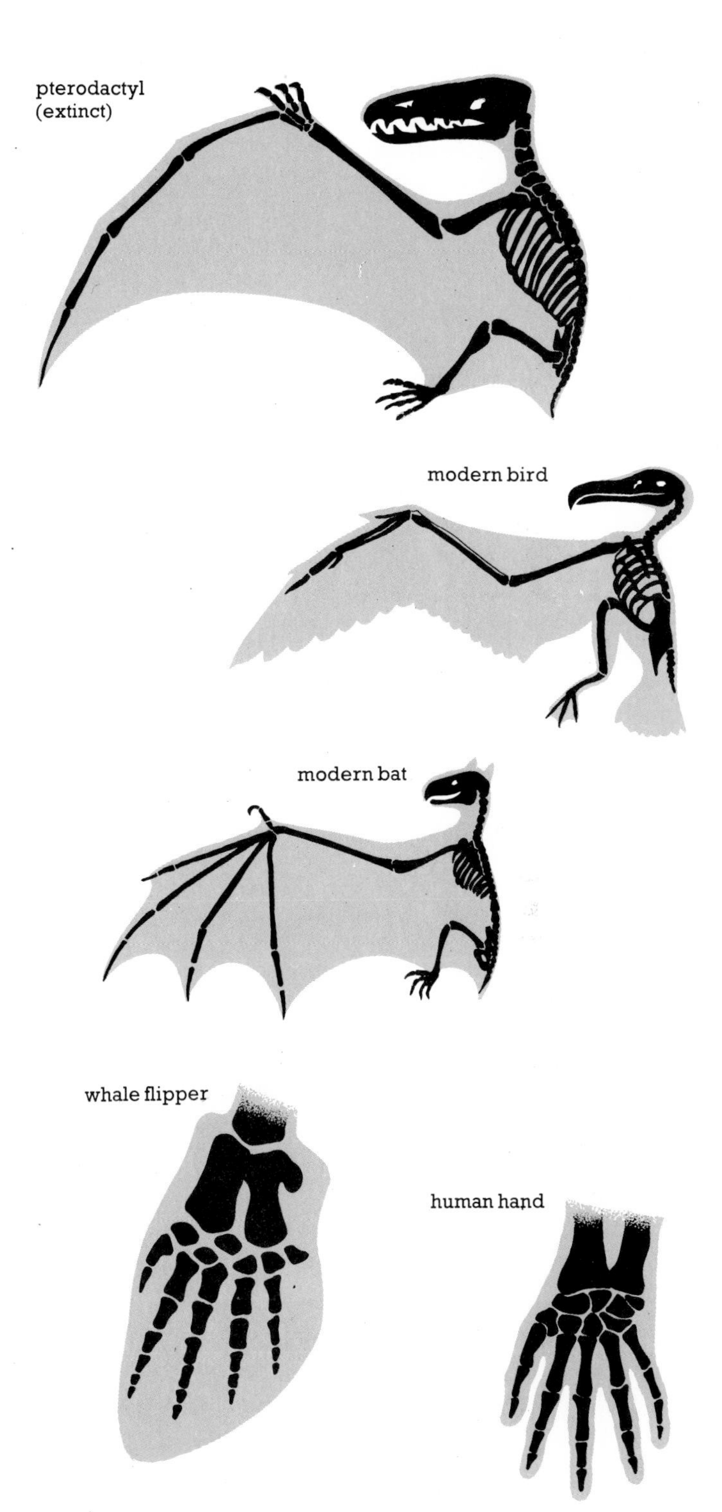

The limbs of different animals (above) show resemblances that indicate descent from a related organ in some common ancestor. All have a basically similar skeletal structure and all occur in a similar way within the developing embryo.

Genes and Evolution

The rules of evolution by natural selection are simple: fit the surroundings and survive; survive and pass on hereditary advantages to offspring. Though these rules are simple, the massive timescale of evolution means that the number and complexity of solutions to the problem of survival have been enormous. Of course, evolution is still continuing, but the process is so slow that scientists usually study evolutionary developments that have occurred in the past.

On rare occasions, however, it is possible to see evolution occurring on a shorter timescale, almost visibly. One of the best examples comes partly from the hobby of collecting butterflies and moths, which was particularly popular in Western Europe and the United States in the nineteenth century. A favorite of the collectors was the peppered moth, which is usually light-colored with darker speckles. But in 1848 a mutant variety was caught near Manchester, in what was then the industrial heartland of Victorian England. This mutant moth was much darker, in fact, almost totally black.

Over the next fifty years or so the dark form became increasingly common in regions of the country that were heavily industrialized. In such regions the original lighter form became rare, but in less industrialized areas of the country the light moth was still the more common.

What caused the evolution by which the dark moth appeared to be replacing its lighter relative? It seems that as areas became more industrialized, increasing amounts of soot and grime from factory chimneys were covering tree trunks and killing light-colored lichens on the bark. Peppered moths have always tended to rest on tree trunks. Their

The light-colored peppered moth (below left) became easy prey for birds when industrial pollution blackened the bark of the trees against which it was previously camouflaged. But the changed environment favored the darker, mutant form (below right) which increased in numbers as a result.

spotted patterns normally helped camouflage them from hungry birds. But on dark tree trunks they were obvious targets, and so were soon eaten. The rare black mutant survived better on the darkened bark, and thus bred and passed on its black characteristic.

The idea was confirmed in the 1950s by the British naturalist Bernard Kettlewell, who watched birds devour light and dark moths that had been deliberately placed on opposite-colored trunks. But the advantage may be passing back to the lighter moths – again due to natural selection. In 1956 a Clean Air Act was introduced in Britain. It was the first such law in the world to create smokeless zones in industrial areas. Because the trees are now clean in these zones, the light-colored form of peppered moth is gradually making a comeback, and the black form is obviously at a disadvantage.

As Darwin himself realized, there was a crucial gap in his theory. He had no satisfactory way of explaining how organisms can accurately pass on newly acquired characteristics. Nor could he explain how, in apparent contradiction, occasional changes occurred that could in turn be precisely inherited. However, our present knowledge allows us to restate Darwin's theory in genetic terms.

Inheritance of characteristics depends on genes. The accuracy of the process is guaranteed by the structure of DNA and the way in which it is reproduced and shared out to sex cells. The occasional mutations are the raw material of evolution. Natural selection favors individuals with certain mutations. But natural selection also dooms those carrying harmful mutations to be among the also-rans of evolutionary history. So natural selection leads to an increase in the percentage of certain organisms, and hence the percentage of certain genes, in a particular species. In fact, in modern terms, evolution can be regarded as a change in gene frequency in a population of organisms.

Demonstrating remarkable ways of hiding from or warning off predators, the leaf insect from Java (above right) is barely distinguishable from the foliage that forms its characteristic habitat; the shape and color of the harmless moth Sesia apiformis *(right) mimics a far from harmless wasp.*

Population Genetics

Although natural selection is probably the main cause of changes in gene frequency, and therefore of evolution, it is not the only one. Evolution can result from several factors operating at the same time on populations of organisms. To determine these factors, studies of gene frequencies are undertaken in a field of research known as population genetics.

The first problem for a population geneticist is to find a population with some easily detectable genetic characteristics. Generally, a geneticist can only manage to study one or a few characteristics at a time. The geneticist then records the frequency in the population of alternative forms of the same characteristic: for example, normal or sickle-cell hemoglobin, curly or straight wings in fruit flies.

Although one obvious and necessary contributor to evolution is mutation, it would take a very long time for a particular mutation to recur often enough to make a substantial impact on a population.

A large number of studies of human blood groups have revealed that gene frequency varies between geographically separate populations of the same species. Why such geographical differences exist is not always clear. One possibility is that gene frequencies are affected by migration. If migration occurs and the immigrant group breeds with a new population, it may cause changes in the frequency of genes in the local population. The extent of the change in the local population would depend on the rate of migration and interbreeding, and the degree of initial difference in gene frequency between the immigrants and locals. Migration can therefore lead to gene flow – a movement of genes from one population into another.

A further factor causing changes in gene frequency depends on the laws of chance. Suppose a gene can exist in two alternative forms and neither form gives any advantage or disadvantage to an individual. Now suppose also that in a large population of organisms each form of the gene has a particular frequency. If individuals mix and breed freely with each other, the frequencies will remain the same from generation

If a rodent population containing a mixture of alternative forms of a gene leading to brown or white fur (a) is split by an earthquake (b), the two main populations may develop different gene frequencies. This becomes more marked (c) if the lone white rodent dies without reproducing.

a

population with genes for brown or white fur

b

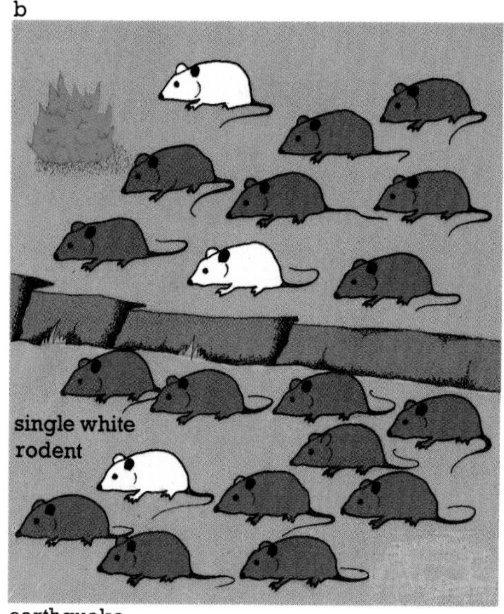

earthquake

c

single white rodent dies without reproducing

to generation.

But what if the populations is small? Here the laws of chance come into play in a strange way. Take the situation where there are ten individuals, and just one has a different form of the gene from the other nine. If that individual happens to die before it breeds, its form of the gene will be eliminated from the population, causing a definite decrease in gene frequency! Or consider a population of several thousand individuals, where only one in a hundred has a mutant gene. This one per cent frequency will remain constant in successive generations. But suppose that a few individuals, say five or six, wander off and form their own isolated population. If, by chance, one or more of these individuals happens to carry the mutant gene, and breeds, the gene may become relatively more common in the new population. The gene frequency will have increased in this population compared with the ancestral one.

As the world map below indicates, the frequency of blood group A varies dramatically. In the light gray areas, the frequency is zero to twenty per cent; in the dark gray areas, it is twenty to thirty-five per cent or more. Frequency is high, for example, among the Blackfoot Indians of North America (above), although it is not yet known why this should be so.

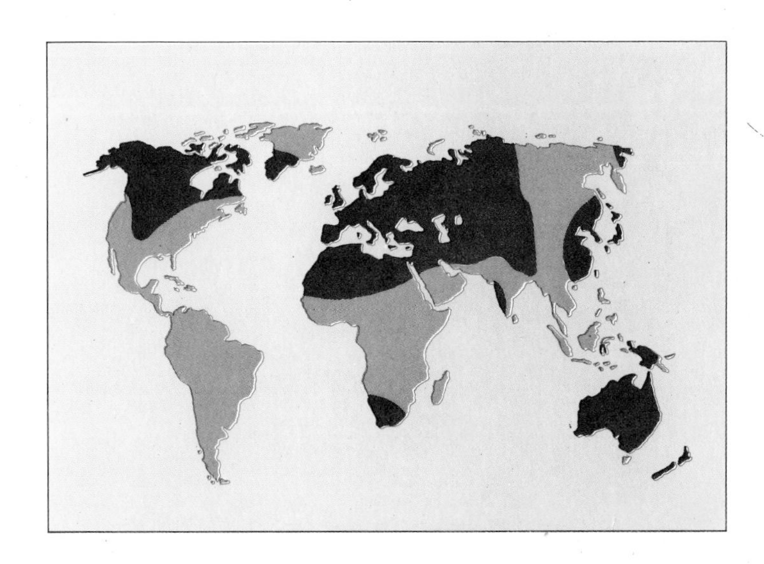

The Origin of Species

Evolution seems to depend on the accumulation of genetic changes that alter characteristics within species (defined as organisms that can breed with one another and produce fertile offspring).

One way in which a single species might evolve into two different ones is by geographical isolation. Some members of a species may leave their group and found a new colony elsewhere as when birds migrate to a new area, or plant seeds drift on the wind. A population may become divided by a major new geographical barrier, for example, a new stretch of water caused by a rise in sea level or the gradual drifting apart of two land masses.

In such isolation, the members of separate colonies can then breed only within their own colony. This may gradually lead to a divergence of their gene frequencies as new mutations arise or different local environments create different pressures of natural selection. Given time, the separate colonies may become so different that they could not breed together even if they were no longer separated. For example, they may evolve incompatible courtship displays or the offspring may be sterile: the chromosomes from the male and female may be too dissimilar to pair up during meiosis.

The differences between species may be clearly visible – as in the appearance of honey bees and bumble bees, for example. Conversely, differences between members of a species can be great – as in the case of dogs; but a collie can breed with a German Shepherd and produce fertile offspring.

In trying to pin down the meaning of such differences, some scientists have come up with more and more sophisticated methods of examining enzymes and other proteins in the cell. Such a substance, when extracted from the cell of a chimpanzee, for example, may be broadly similar to a protein fulfilling a comparable function in a man. But the proteins may differ in detail, that is, in the precise sequence of their amino acids. These differences are due to differences between the sequences of bases in the DNA of the gene that determines the protein in each organism.

Using the genetic code, it is possible to estimate the

The animal above is the result of a successful attempt to crossbreed a zebra with a horse. The tigon (below) is a hybrid between a lion and a tiger. Both animals highlight the close relationship between their respective species. However, the evolution of zebras and horses, and lions and tigers, have followed separate paths for so long that the hybrids are unable to produce any offspring of their own.

number of mutations theoretically required to go from the amino acid sequence of one protein to that of another. These numbers may then be applied to draw up a family tree of proteins. Small numbers of intermediate mutations mean minor differences between proteins, which are therefore placed close together on the tree. The result largely resembles the tree drawn up by the older method of comparing visible large-scale features of whole organisms: in both, a human is placed close to a chimpanzee but a long way from a tuna fish or a frog.

However, this technique presents some problems. The proteins from a chimpanzee and a human are so similar that we might wonder why the organisms are, in fact, so different. The answer may be that some new species do not arise by the accumulation of small changes, but by more dramatic jumps.

Darwin found each of these species of finch (shown above, with male and female of each species) in a different environment in the Galápagos Islands, which lie 600 miles off the coast of Ecuador. Though in many ways alike, each species seems to have evolved so as to cope successfully with its particular surroundings. The main difference between the birds is their beaks, the size and shape of which are related to the birds' diet. For example, the staple food of the cactus ground-finch (Geospiza scandens)*, top left, is the flowers of the prickly pear, so its long beak is admirably suited to flower-probing and nectar-feeding. Whereas the medium ground-finch* (Geospiza fortis)*, bottom right, feeds chiefly on large seeds, hence its huge, heavy beak.*

Behavioral Genetics

An animal's behavior helps it to cope with its surroundings – for example, to find food, attract a mate, and rear its young. Patterns of behavior are often very complex as shown by, for example, courting displays in birds of paradise; stalking of prey by lions and cheetahs; hive building by bees; rearing of young by monkeys; and, of course, the most varied and complex of all behavior, human activity.

But are there genes that govern complex patterns of behavior? If successful behavior has a genetic basis, presumably a successful animal would breed and pass such genes on to further generations, which would, in turn, have a higher chance of survival and of reproducing because of those genes. Natural selection would strongly favour the evolution of such genes. But do they actually exist? That question is very difficult to answer.

Patterns of behavior have been studied in detail for many years and have proved to be very complex. For example, ants are often seen to carry away the corpse of a dead ant and dispose of it away from the nest. The live ants are able to recognize the dead ant by its smell. Experimenters have isolated the chemical producing this smell, and smeared it on the body of a live ant. Despite its obvious signs of life, the smeared ant is promptly carried off as if it were dead! Ants respond to the smell in a very fixed way. They ignore other signs that indicate that the smell is misleading.

It is likely that the rigid behavior of ants is inherited, not learned, and the same can be said for many other similar sorts of behavior. However, the explanation of such recognition mechanisms is often more complex. Many newly hatched birds will, as soon as they are able, follow their mother. Is this recognition of the mother inherited? The Austrian biologist, Konrad Lorenz, one of the founders of modern behavioral studies, showed that newly hatched geese would

The elaborate courtship display of the greater birds of paradise, from New Guinea, is a striking example of the complex patterns of behavior to be found among animals.

These two young bull elephant seals are fighting each other at the start of the mating season. The winner of the combat will mate with the harem of females nearby. The courtship and mating ritual among seals is a behavioral pattern that has been closely studied by scientists.

follow the first thing they encountered, whether it was their mother, another bird, a mechanical toy, or even Lorenz himself. Once followed, the object became "mother" to these birds. The ability to recognize the real mother was therefore not instinctive – that is, not inherited. What seemed to be inherited was the tendency to follow the first available object that could possibly be regarded as a mother. Yet even this apparently instinctive activity may depend on the chick recognizing sounds it has already heard while still in the egg; sounds of its mother, or perhaps of Lorenz imitating the call of a bird. Of course, in the wild this limited inherited behavior would serve its purpose, because the real mother is likely to be the one heard in the egg, and the first object to be seen on hatching.

Behavior patterns that are often regarded as inherited, such as mating, or even so-called maternal instinct, are not necessarily inherited as such, but may instead be learned. Female monkeys deprived of their mothers as babies often show a much reduced capacity later to care for their own young. This suggests that monkeys acquire their maternal ability by imitating their own mother.

Just how important genes are in animal behavior is currently a highly controversial area of biology. Generally, it seems that the higher an animal is placed on the evolutionary tree, the less one can regard complex behavior as purely inherited. What higher animals inherit is probably the capacity to learn, plus some relatively simple instinctive responses that protect them, particularly when young.

Animal Breeding

People have kept animals for food, clothing, and transport for at least 10,000 years. Over much of this period they have tried to breed them selectively. There have been two aims. First, when, by chance, a useful or interesting variety of animal has turned up, people have mated it with its own kind, for example, its close relatives, in an attempt to retain the desired characteristics in future generations. This process, known as inbreeding, is still widely practiced today, particularly with domestic animals, such as cats and dogs, which are bred for their appearance.

The opposite of inbreeding is outbreeding. This has been practiced when people have tried to improve existing varieties by breeding them with other varieties whose characteristics are considered to be desirable, again with the intention of improving future generations. This outbreeding has been commonly practiced by farmers, for example, to improve beef quality in cattle, or wool in sheep. Racehorse breeders also try to outbreed, say by crossing a fast mare with a strong stallion. Such techniques have been used for centuries, long before breeders had any knowledge of genetics. So what difference does present-day knowledge make?

Genetic knowledge can help breeders design programs aimed at producing desired varieties. We know that genes and the environment interact, and it is important to be sure that a desirable characteristic is dictated by genes before embarking on an extensive inbreeding program. For example, egg production

The Australian merino sheep on the left is the result of nearly two centuries of selective breeding, following its introduction to Australia in 1796. Another merino on the right has been bred to retain the characteristics of the ancestral Spanish merino sheep.

Boxer dogs (left) are the descendants of numerous breeds of the bulldog type. In the nineteenth century German breeders developed the modern form shown here. Boxers are alert and intelligent, but undesirable characteristics can be the price of inbreeding. Boxers often develop respiratory and dental problems owing to their short noses and jaws.

in chickens seems to depend largely on environment. Farmers wanting to increase egg production would therefore achieve less by selective breeding than by providing the most suitable environment.

Breeders also now know that characteristics depend on the particular form of genes. To produce true-breeding varieties, that is, varieties that consistently exhibit certain characteristics from generation to generation, one must obtain individuals in which both genes of the gene pair determine those characteristics. This can be achieved by inbreeding, but that approach may also lead to an accumulation of harmful recessive genes, which are normally masked by their dominant partners. This results in inbreeding depression, characterized by such defects as stunted growth or feebleness in the inbred animals.

One possible solution to this problem is outbreeding, in which potentially harmful recessive genes may often be compensated for by a dominant form. This dominant form is more likely to be present in a nonrelated mate than in a close–genetically similar–relative. The introduction of new genes into inbred stocks by outbreeding to other stocks makes it harder to achieve consistent qualities, but it can lead to vigorous offspring.

The most extreme examples of outbreeding are achieved by mating individuals of different species.

Mating a horse with a donkey produces the sterile mule (below), once very popular because of its strength and endurance.

This has only rarely been successful in animals; even then the resultant hybrids are usually sterile. (This use of the term "hybrid" should be distinguished from Mendel's hybrids, which were the results of the cross-fertilization of one plant with another of the same species.) However, sterile hybrids can be useful enough to make it worth continuing their production by crossbreeding the parent species. For example, horses and donkeys have been mated for centuries to produce mules, which, though sterile, are nevertheless very useful as beasts of burden.

Plant Breeding

Mankind has selectively bred plants for thousands of years with the aim of improving their quality. Experimenting with cereal crops, breeders have aimed to enhance their yield of grain, the quality of their flour, and their resistance to drought or disease. With other plants, breeders have tried to improve the perfume and color of the flowers.

Improvement of plants, particularly food crops, is obviously important and genetics has contributed to a better understanding of the benefits and disadvantages of particular breeding programs. Many cereal crops such as corn are now planted largely as hybrid seed, produced by outbreeding between different inbred varieties. The vigor of the hybrid plant is probably a major contribution to the increased corn output in the United States. This increased output represents a major achievement for applied genetics. In 1929 practically no hybrid corn was grown among the 100,000,000 acres of corn in the United States. But by 1970 the vast majority of 67,000,000 acres was planted with the hybrid variety, yielding twice as much corn.

Plant breeders have a definite advantage over animal breeders, because they can often produce fertile varieties – indeed, new species – by crossbreeding between species. This is because hybrids are often polyploid. Polyploidy, as we know, can occur naturally in the wild. Some species of cotton that we now grow are polyploids that probably arose originally by accidental crosses between different species of cotton.

But breeders do not have to rely on accidents. They can attempt to produce fertile polyploids by crossbreeding between different species. One early attempt to produce a new hybrid species was made in 1927 by the Russian geneticist G. D. Karpechenko, who crossbred two quite distantly related species, a radish and a cabbage. Each species has eighteen chromosomes (nine pairs); the hybrids had the same number (nine radish chromosomes and nine cabbage chromosomes) and were sterile. However, some polyploids arose by chance. These had thirty-six chromosomes (nine pairs of radish, and nine pairs of cabbage), and were fertile. Unfortunately, the hybrid was not commercially successful because as luck would have it, the plant had the leaves of a radish and the roots of a cabbage!

Breeders can artificially encourage polyploidy by treating the hybrids that result from crossbreeding between species with a chemical called colchicine obtained from autumn crocuses. This chemical allows the chromosomes to reproduce, but prevents

Hybrid vigor and the damaging effects of inbreeding are demonstrated by these corn plants. The parents are on the left and the vigorous hybrid offspring is third from the left. Next, descending in height, are successive generations produced from the hybrid by inbreeding.

the formation of two separate cells. The number of chromosomes in the nucleus is therefore doubled. Several of these new polyploid varieties promise to be very useful. For example, a new hybrid cereal, called triticale, produced by crossbreeding rye with a species of wheat, adds rye's resistance to cold winters to the usual properties of wheat.

Recent breeding programs have led to highly inbred wheats. Much of the genetic variability, that accumulated over 9000 years of wheat cultivation, is missing from present-day varieties. If a new disease should arise, or if the climate were to change suddenly, much of the wheat might be damaged and lost. So it is a good idea to introduce other genes into wheat by outbreeding. One way of doing this is to crossbreed the inbred varieties with their wild relatives, which may be resistant to viruses, insects, or drought. For this reason, some wheat breeders believe it is essential to conserve some stocks of primitive wheat in seed banks, from which they will be able to take a transfusion of genes, if and when the need arises.

In the field of primitive emmer wheat (above left), seen growing in northern Israel, individual plants show wide variations in color and height. In contrast, the field of modern bread wheat (left) – one of over 20,000 cultivated varieties – growing in the same area, shows remarkable consistency between plants.

Varieties of the same species of the cabbage family have been selectively cultivated for their differing features of taste and appearance. Some varieties became hard-headed, like a modern cabbage (a); some made masses of flower buds, as in cauliflower (b) and broccoli (c); and some made clusters of leaf buds, as in Brussels sprouts (d).

Cloning and Cell Fusion

A major problem in studying inheritance in animals or plants is time. Often weeks, months, and years elapse before offspring can be produced and their characteristics examined. In humans, especially, the long gap between generations and the impossibility of doing actual breeding experiments make it difficult to find out about genetics. As a result, scientists have used organisms that breed quickly in order to unravel the complexities of genetics. Fortunately, the principles they discover generally hold true for other organisms. New techniques have now revolutionized our knowledge of human genetics, and promise to speed up advances in plant, and even animal breeding, too.

Geneticists can now take the somatic cells of plants, such as tobacco, geraniums, and carrots, and grow them in tissue culture. They can then take several, genetically identical, cells from the culture and get each to grow into a whole plant. Such plants are also genetically identical. This technique is therefore a way of propagating several plants, with desirable genetic characteristics, from an initial single plant, without the need for an inbreeding program. Producing identical plants in this way is called cloning.

Geneticists have not yet succeeded in doing precisely the same with animals. However, animal cloning can be achieved in a somewhat different way. Nuclei taken from identical tadpole cells in tissue culture can be injected into a series of eggs whose own nuclei have been killed. The eggs then develop into a group of genetically identical cloned adults. Such cloning is in its very early stages, but it could be a useful technique for the future. For example, a cow with a record-breaking high milk yield could be cloned to produce a whole herd of identical cows.

Another way of manipulating somatic cells has already helped to increase our knowledge of human genetics. Two or more cells may occasionally fuse together to produce a single cell. This fusion can be induced artificially by adding certain chemicals or

This diagram shows how the nucleus of a tadpole cell can be inserted into a frog's egg, whose own nucleus has been killed by ultraviolet radiation. Development occurs, and eventually the egg gives rise to a complete tadpole. Many eggs can be treated in this way, using genetically identical nuclei from the same tissue culture, thereby producing cloned tadpoles.

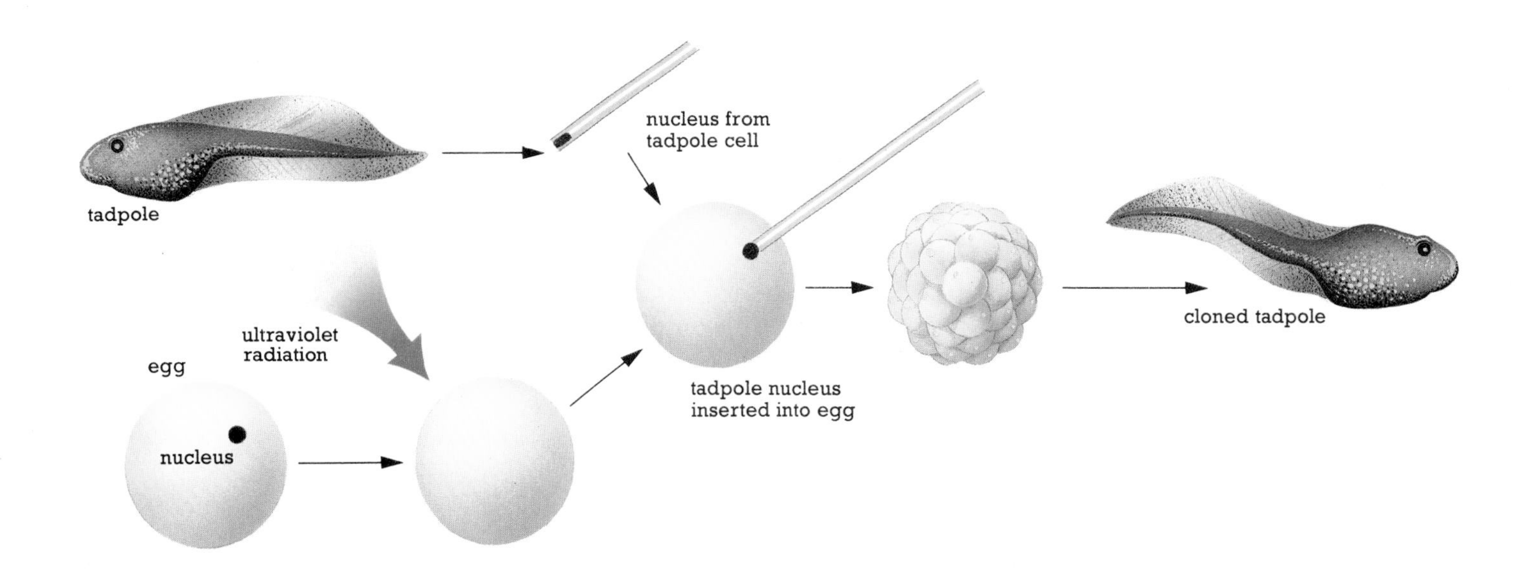

The large eggs of the African clawed frog, Xenopus laevis *(left), are widely used to investigate the ability of nuclei from other, somatic, cells to support normal development of another frog. This is called propagation by cloning.*

viruses. In a series of experiments conducted during the mid-1960s, scientists working at the laboratory of the British geneticist, Henry Harris, in Oxford, England, succeeded in fusing cells taken from a variety of species, including chickens, frogs, mice and humans. The hybrid cells that resulted were of two types. Some contained two or more separate nuclei, while the nuclei of others fused to give a giant nucleus containing all the chromosomes. For example, if a mouse cell was fused with a human one, the result was a single nucleus in which mouse and human chromosomes could be distinguished by their appearance. This dual set of mouse-human chromosomes could be copied and passed on to the daughter cells by mitosis. However, the reproduction process is not very stable and during successive stages, chromosomes tend to be lost. The process appears to be selective and particular chromosomes vanish over successive generations of cell division.

The gradual loss of chromosomes from dual sets can be used by geneticists as a clue to which particular chromosomes carry certain genes. In mouse-human hybrid cells the human chromosomes are lost rather than the mouse ones. By noting the presence or absence of characteristics – for example, particular enzymes – as chromosomes are lost, geneticists can establish which chromosomes carry the genes for those characteristics. If geneticists follow several characteristics at once, they can see if the loss occurs with the loss of the same or different chromosomes; in other words, they can tell whether the genes of those characteristics are linked. By following a few rapidly breeding generations of hybrid cells, geneticists are able to map genes in humans in a way that would take perhaps hundreds of years by the traditional methods of human genetics.

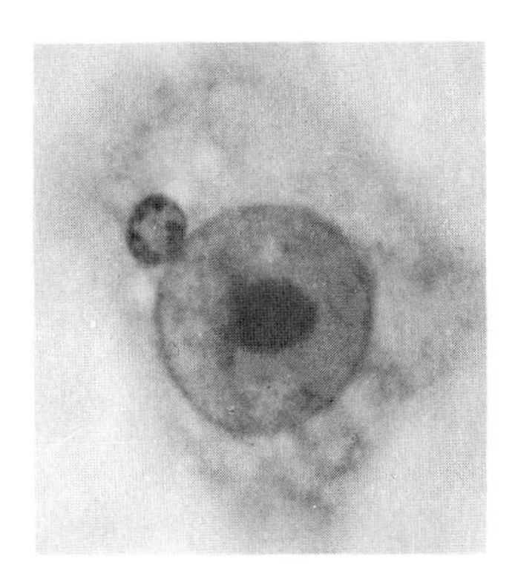

This hybrid cell contains a large, human nucleus and a much smaller nucleus from a chicken. A virus was used to bring about the cell fusion. However, no such hybrid has yet grown beyond the cell stage.

Human Genetics

A human being is just one organism among many. But, not surprisingly, humans have received a great deal of attention from geneticists. It was probably through comparing themselves with their relatives that humans first formed ideas about inheritance. The philosophers of Ancient Greece were aware that certain things, such as blue eyes or baldness, were probably inherited, but their view as to how this occurred was far from accurate.

Present knowledge about human genetics still depends to a great extent on studying humans themselves. For example, by tracing the inheritance of certain characteristics in families, scientists have discovered that a large number of such characteristics depend on genes. Thanks to well-documented characteristics, such as hemophilia, geneticists do have access to some long-term information. In this century intensive studies of large numbers of families have resulted in many additions to the list of known human inherited characteristics. For instance, there are some 2300 human diseases that are now known to be inherited, and the list is constantly growing. In some cases scientists also know whether inheritance is due to a dominant or recessive gene, and whether that gene is sex-linked.

Knowledge of human genetics has another important implication for medicine: an awareness of "invisible" genetic differences between individuals can put doctors on the lookout for unforeseen reactions to certain drugs. For example, about two per cent of people have been found to react abnormally to the drug succinylcholine, which is used during surgery to relax muscles, including those that normally help us to breathe. As a result, patients must be helped to breathe artificially during the operation. Normally, the effects of this drug last for only a few minutes because it is rapidly broken down by an enzyme in the blood plasma called cholinesterase. But in some people cholinesterase works more slowly, so that the effects of succinylcholine persist much longer. As a result, patients may not begin breathing until several hours after the end of an operation. The lack of oxygen can obviously be fatal unless doctors are forewarned. Researchers have now found that slow-acting cholinesterase runs in families, that is, it depends on a genetic variation. Careful research before an operation can therefore enable doctors to avoid fatalities in affected people.

The discovery that the basic genetic mechanisms in humans are like those in all other living organisms suggests that humans, like them, evolved from some single-celled creature of about 3.5 billion years

Despite their greatly different heights and facial features, these two women are twin sisters. The genetic differences between non-identical twins are likely to be as great as those between any two offspring of the same parents.

Gout is a condition that causes attacks of acute arthritis in middle-aged males. It may become chronic and deforming. It was once thought that gout was caused by excessive consumption of alcohol. Eighteenth-century caricatures of elderly alcoholics often included a heavily bandaged, swollen foot to represent the fact that they had the disease. Scientists now believe that some forms of gout may be inherited.

ago. The later stages of this evolution, from early apelike man to our present state, probably took a mere four million years. All these processes involved changes in gene frequency.

But what of more recent human evolution, in which mankind has domesticated animals, planted crops, built homes, used fire, invented tools and the wheel, written poetry and music, fought wars, and traveled to the moon? Is this development also due to changes in gene frequency? The likely answer is no! What has happened is that humans have developed a complex culture, which they can transmit by using language and that in turn shapes their own environment. This cultural inheritance is essentially independent of genes, except in the obvious sense that, without genes, no human beings would exist, and their genes provide them with a capability to create their culture. Cultural changes now occur at a speed that far outstrips biological evolution and can make evolution of less importance when considering future human society.

Possession of an attached earlobe (left) is a recessive characteristic that may appear in families every few generations. The unattached earlobe at right is a dominant characteristic.

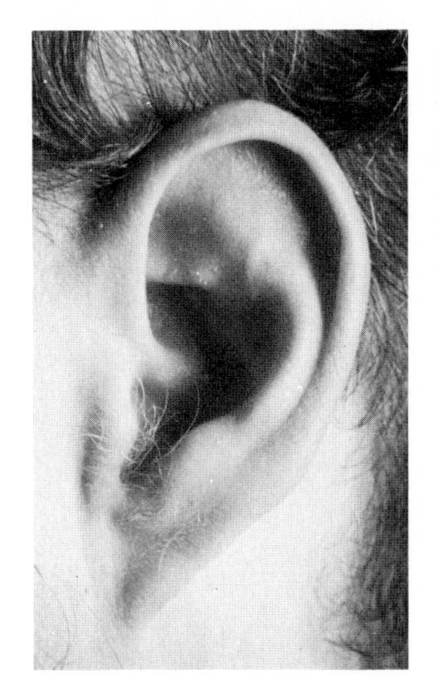

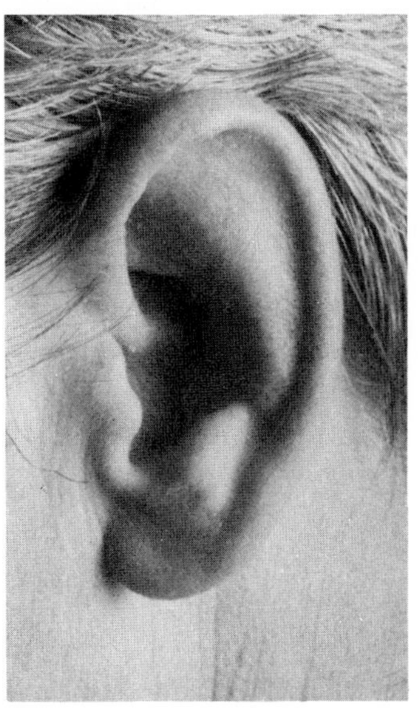

Genetic Counseling

Most people know of certain characteristics that are shared by several members of the same family – blue eyes, red hair, and so on – without considering whether such characteristics are passed on by genes. Most of the time this does not really matter. But occasionally the question of inheritance becomes very important. For example, if a young man and a woman are planning to get married, and the woman discovers that two of the man's cousins and one of his grandparents suffer from some unpleasant disease, she might have doubts about whether she should marry him. Would their children have the disease; is it genetically inherited? On such occasions, people sometimes seek the advice of a doctor, who may in turn refer them to a genetic counseling service.

But even if the counselor diagnoses a disease as genetic, this does not mean that the couple's children will inherit it. If, for example, the disease is caused by a recessive mutation, then their children would develop it only if they lack the dominant, normal form of the gene. Since, in this example, the man had not got the disease, at worst he might have one normal and one mutant gene. So there would be only a one in two chance that he would pass on the mutant gene if he had it. Even if this did occur, the mutant would most likely be compensated for by a normal gene from the mother. However, in other instances, there might be a likelihood of a genetic disease being passed on. For example, it might occur in both families, the man's and the woman's. Having considered the counselor's advice, the couple can then decide for themselves whether or not to have children. But suppose a woman is already pregnant, and has reason to believe that her child will be born severely abnormal? Can anything still be done?

Fortunately, the answer is yes. Doctors can now diagnose whether certain incurable genetic diseases are present in the developing child by studying samples of cells taken from the fluid surrounding the baby while it is still in the womb. If signs of the disease are found, the couple may then decide to have an abortion.

Occasionally social or political groups are formed that are dedicated to improving the human race by genetic means. This use of genetics, which is sometimes called eugenics, might involve avoiding undesirable genetic diseases by preventing, or dissuading, breeding between some individuals. It might also encourage marriage and breeding between others thought to have desirable characteristics. Eugenics is not widely popular – for several good reasons. Often it is hard to be sure whether characteristics are genetically inherited. Even if they are, in the case of some diseases, it is impossible to tell how severe they are likely to be. What is desirable or undesirable is often simply a matter of opinion, with which not everyone might agree.

There are other problems. Ellen Terry, the beautiful English actress, once invited the famous Irish wit and playwright, George Bernard Shaw, to have children with her. She thought that with his brains and her beauty any children would be truly wonderful. Shaw pointed out that their children would be just as likely to have his beauty and her brains, a very undesirable combination! As with attempts to crossbreed radishes and cabbages, the trouble with selective breeding in humans lies in not getting what is planned.

Indeed, geneticists themselves are far from certain whether characteristics such as intelligence are inherited biologically. In attempting to "improve" the human race, it is a mistake to look only at genetics. Human characteristics depend on interactions between genes and environment. A feature might therefore, just as easily, or indeed more easily, be altered by altering the environment as by preventing or encouraging inheritance of particular genes.

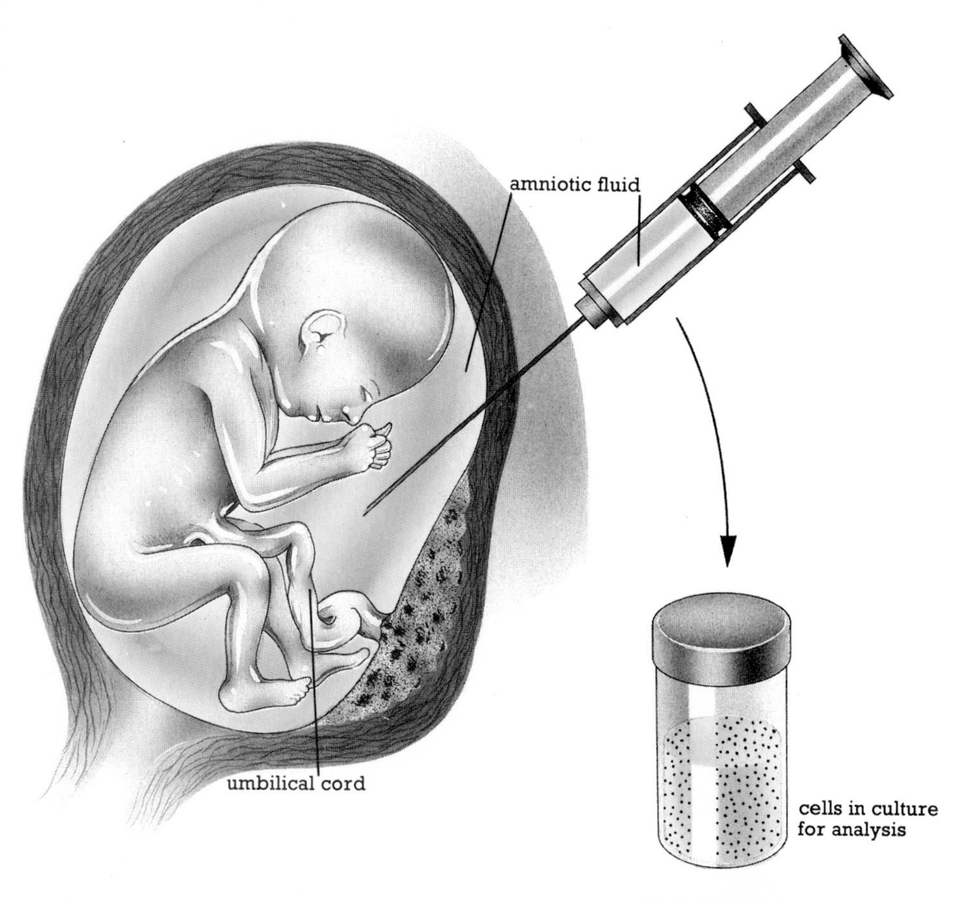

The diagram at left shows how a sample of cells can be safely removed from the amniotic fluid surrounding the fetus. The chromosomes of these cells can then be analyzed and checked for any of the fifty abnormalities that can now be detected through this technique.

An image of a human fetus in the womb, visible on the monitors, can be obtained by reflecting sound waves through the mother's abdomen. This technique is safer than X-rays and allows doctors to establish the exact position of the fetus.

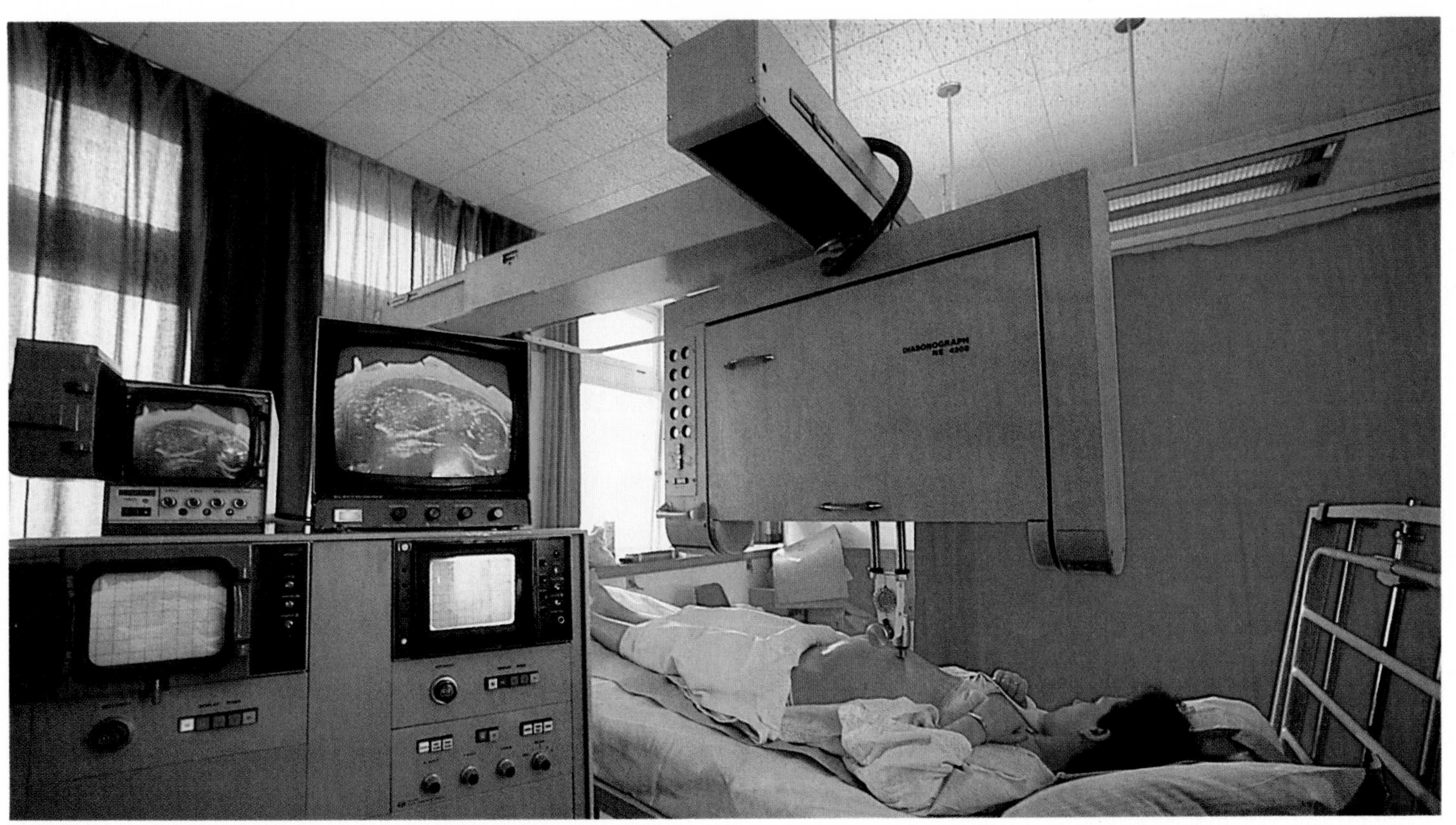

Genetic Engineering

Genetic engineering, one of the most exciting developments of modern science, involves a range of techniques that allow geneticists to transfer genes from one organism to another. In principle geneticists use simple, rapidly reproducing organisms, such as bacteria, as chemical factories for manufacturing products needed by other organisms, such as human beings. One such product is insulin.

Many people suffer from a disease called diabetes, which results from their insufficient production of insulin. In healthy individuals insulin is produced by an organ called the pancreas, located near the small intestine. Insulin acts as a chemical messenger or hormone, and instructs the tissues to absorb glucose from the blood. To make up for their deficiency of insulin, some diabetics can be given supplies of the hormone from animals. But animal insulin differs slightly from the human variety and sometimes has unpleasant side effects.

Through genetic engineering, however, scientists have now developed techniques for transferring the gene containing the instruction for insulin from human cells to the bacterium *E. coli*. The bacterium reproduces very rapidly and successive generations of its descendants' cells contain the human insulin gene. These generations of bacteria can accurately read the instructions in this gene, and therefore produce fresh supplies of human insulin. Bacteria can be grown in huge amounts, in large vats similar to those used in brewing beer; and, therefore, the use of human insulin produced by bacteria should be common within a few years.

Similar genetic engineering techniques can be used to create new varieties of bacteria producing other substances, which are also in short supply. A number of such projects are already under way in the United States, Japan and Britain. Some projects aim to produce hormones; others to create vaccines against infectious diseases.

Genetic engineers also strive to create new bacteria for other purposes. For example, there are

In the diagram below, DNA from the cells of a plant or animal (a) can be extracted (b) and broken down by means of enzymes (c). Geneticists then splice each fragment into a circular carrier DNA molecule (d and e). These circles are inserted into bacteria (f). The bacterium carrying the required gene (DNA fragment) is then selected (g) and allowed to reproduce. All members of the colony from this selected bacterium will carry the required gene (h).

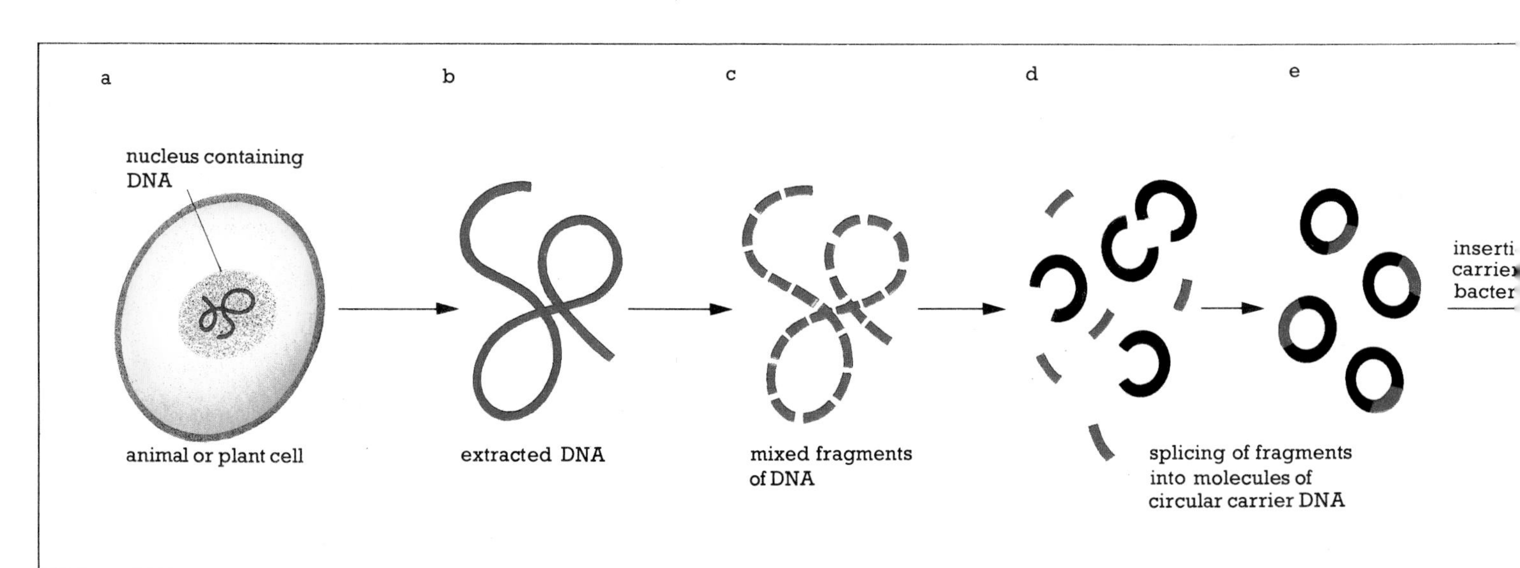

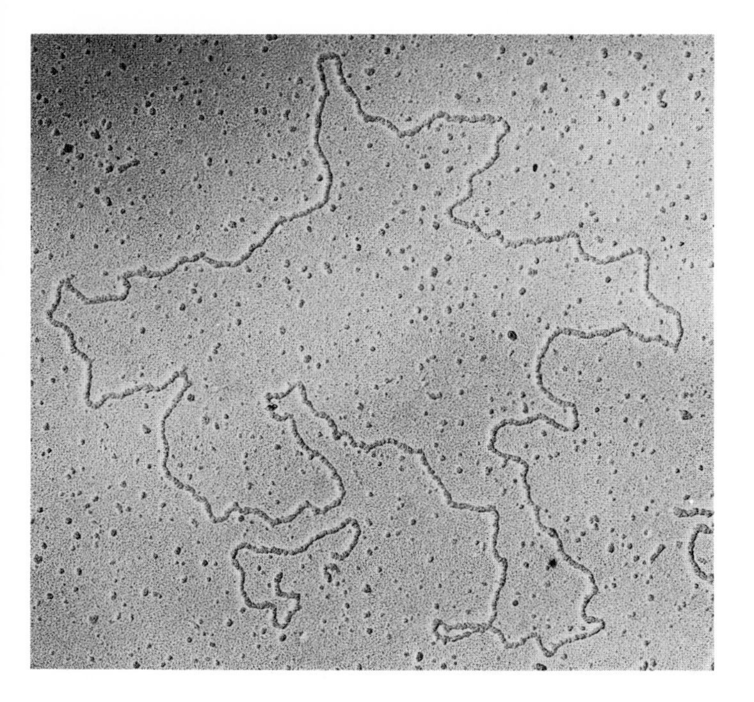

This electron micrograph shows a circle of DNA of the type used as carriers in genetic engineering. This particular one comes from the bacterium Streptomyces coelicolor.

plans to produce bacteria that will mop up oil slicks by digesting the oil. Nor need bacteria be the only host organisms for genes. Scientists are attempting to isolate genes that enable certain bacteria to trap nitrogen from the air, and transfer these genes into edible plants, particularly cereals, which are unable to obtain nitrogen in this way.

A number of scientists working in Israel and Italy have already attempted to cure an inherited disease, called thalassemia, by genetic engineering. This fatal disease results from the production of abnormal hemoglobin. The scientists attempted to transfer genes for normal hemoglobin to bone marrow, where red blood cells are made. Their attempts appear to have been unsuccessful, but some day successful treatment may be commonplace.

Though genetic engineering holds great promise, the problems that remain are twofold. First, are we justified in altering the genetic structure of any organism, including humans, in this way? Second, can this be done safely and effectively? Pessimists fear that hidden dangers may arise. For example, imagine a virus that had been deliberately designed to stimulate human cells into growth, perhaps to help tissue repair. This same virus might lead to an epidemic of malignancy if its application were not very carefully monitored.

On the other hand, optimists consider how society may benefit in the future. Imagine introducing new genes into human cells so that we would no longer require vitamins and could make all amino acids ourselves, or could digest cellulose.

The challenge is to harness the results of genetic engineering experiments so that they are acceptable, effective – and safe – for tomorrow's world.

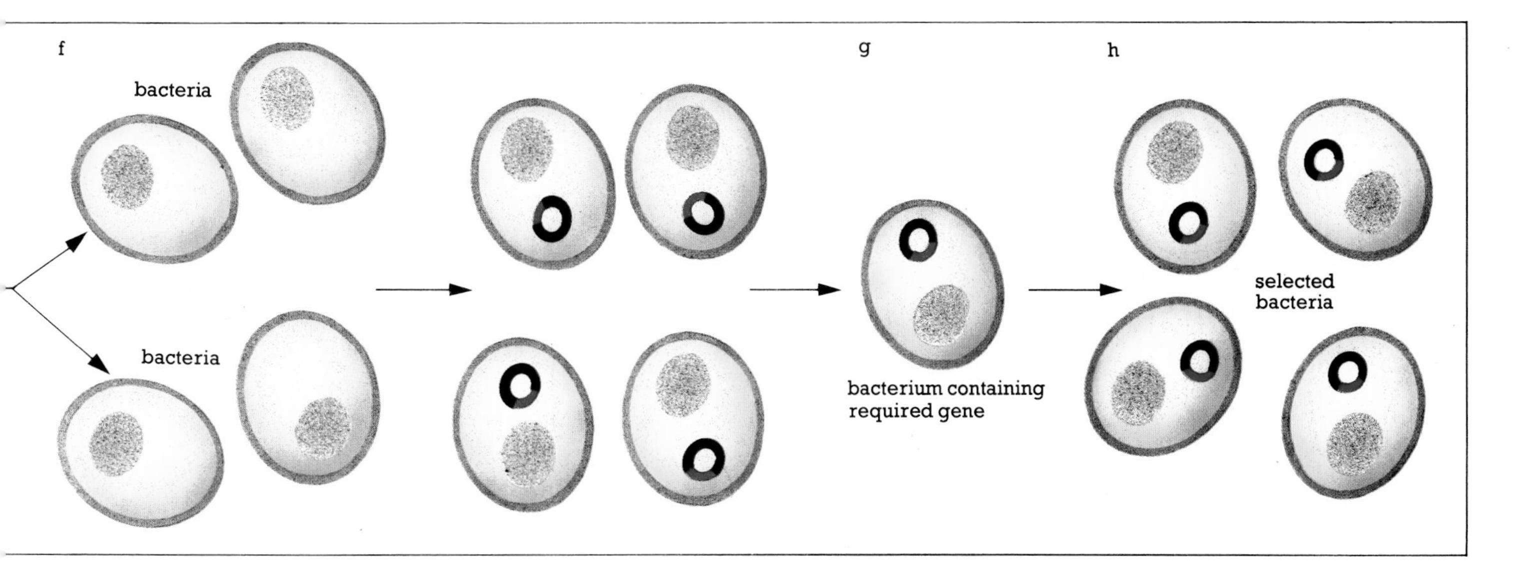

Genetics and the Future

Recent advances in genetics have been very impressive. But as well as solving some problems and promising to solve others, these advances have also raised new questions. For example, it now appears that not all the massive amount of DNA found in cells in higher organisms necessarily contains instructions for making proteins. What, then, does it do? The answer is as yet unknown. It may, as some geneticists suggest, have no function at all. For this reason, it is often called "nonsense" or "junk" DNA. Geneticists may, of course, find that what is now called junk has important functions yet to be revealed.

In practical areas, too, such as medicine, food production and alternative energy sources, there is considerable potential for genetic advances. For example, in Brazil, alcohol produced from sugar cane is currently being used as fuel for automobiles. But one day it may be possible to engineer sugar cane genetically so that the cane produces alcohol directly. Advances in gene-transfer techniques may also one day lead to the creation of bacteria able to convert waste products, such as paper or plastic, into useful fuels or raw materials.

What new things will genetics do for human beings? First, increasingly accurate methods of mapping genes should help in identifying the precise causes of a number of inherited diseases and abnormalities, which doctors can then set about preventing or curing. In some cases the cure may itself involve genetic engineering. People lacking particular hormones or enzymes may have the genes for these implanted into their cells. In the future there may be diabetics who can make insulin, and hemophiliacs whose blood clots normally.

But will it be possible to eradicate inherited diseases altogether? There are two approaches to this question. The first involves techniques, which are continually improving, for predicting the characteristics of unborn children. The second approach is to consider ways of altering the genes of the embryo while it is still in the mother's womb.

In a series of experiments at Oxford University, conducted recently by the American molecular biologists Franklin Costantini and Elizabeth Lacy, genes for the beta chain of hemoglobin from a rabbit were injected into mouse eggs. These eggs were then placed in the wombs of mice. Some of the new-born mice that resulted were found to contain rabbit genes in their cells. When some of these new-born mice were subsequently mated with normal untreated female mice they passed on their rabbit genes to their offspring. So it may one day be possible to supply or replace genes at very early stages of human development, and the new genes may be passed on through successive generations.

When considering the future of genetics, it is very easy to get carried away and imagine that further advances in genetic engineering will solve all our problems. But genes do not act alone. They operate within, and interact with, a particular environment. And both genes and environment affect the quality of life. Only by skillful engineering of both genes and environment will it be possible to ensure the future of genetics as a potent force for changing and improving the world we live in.

The photomicrograph opposite shows Rhizobium *bacteria infecting a root hair of a clover plant, a process that benefits both organisms. The bacteria form a thread that extends from the tip of the root hair into the interior of the root, and supplies the plant with fixed nitrogen essential for growth. One day, geneticists may be able to present farmers with genetically altered crop plants that can fix their own nitrogen, thus avoiding the need for expensive nitrogen-containing fertilizers.*

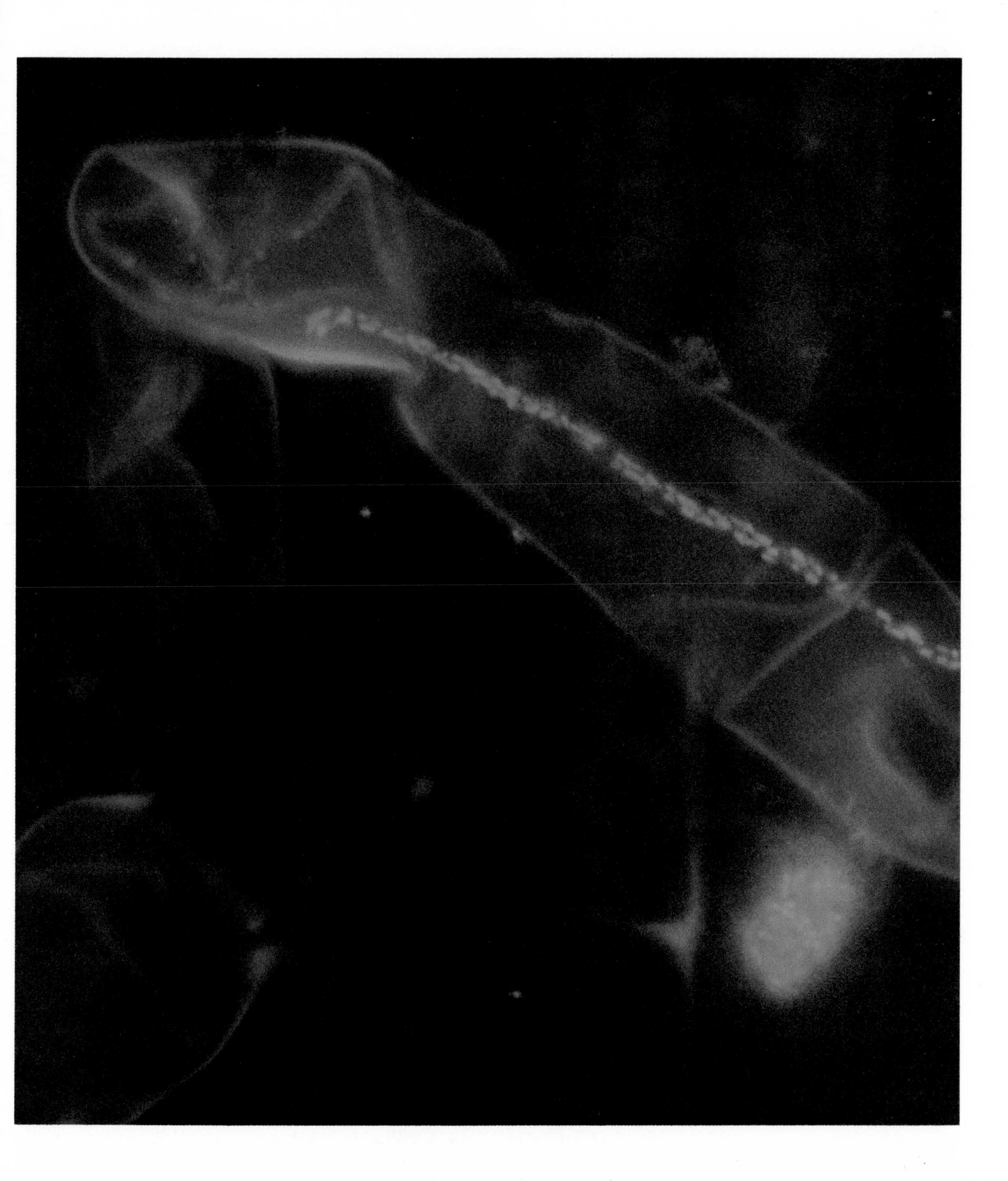

Glossary

Albino: an animal lacking pigment in the skin, hair and eyes

Amino acid: a small molecule that is a building block of proteins; twenty different types of amino acids commonly occur in proteins

Antibody: a specific protein produced in response to the presence of a foreign substance. An animal has the capacity to produce millions of different antibodies

Asexual reproduction: the reproduction of individual cells or organisms that does not involve fertilization

Bacterium: (*plural* bacteria) a simple single-celled organism that has no distinct cell nucleus and reproduces asexually

Base: a building block of DNA. There are four types of base in DNA: adenine (A), cytosine (C), guanine (G) and thymine (T)

Blood group: classification of blood according to types of specific molecules on the surface of the red blood cells; the types and hence the groups are genetically determined

Cell: the smallest self-contained living unit. Some organisms consist of just one cell while others are made up of many millions

Cell differentiation: the production of different types of cell that occurs during development

Cell nucleus: a central structure in cells (except bacteria and blue-green algae) that contains the chromosomes

Cell membrane: the outer layer of a cell that acts as a selective barrier controlling what passes in and out of the cell

Chloroplast: a structure inside a plant cell where the energy in sunlight is trapped and used to help convert carbon dioxide into sugars

Chromosome: a structure in the nucleus of animal and plant cells consisting of DNA (the genes) wrapped up with specific proteins called histones

Cloning: a series of techniques for producing cells and organisms with identical genes

Codon: three consecutive bases in the messenger RNA that are read to specify an amino acid

Cytoplasm: that part of the cell outside the nucleus

Diploid: a cell containing two complete sets of chromosomes, one set deriving from the father and one from the mother

DNA (deoxyribonucleic acid): the substance of which genes are made. It comprises two very long strands of bases wound round each other in a spiral

Dominant: a gene that specifies its characteristic to the exclusion of a corresponding (recessive) gene

Embryo: a developing organism before birth or hatching

Enzyme: a specific protein that speeds up a chemical reaction

Eugenics: the attempt to "improve" the human species by genetic means

Evolution: change leading to a new species of organisms

Fertilization: the fusion of egg and sperm (or pollen) resulting in a diploid fertilized egg

Gene: a particular unit of inheritance; instructions for making a specific enzyme or structural protein; a sequence of bases in DNA

Gene frequency: the proportion of a particular form of a gene occurring in a population

Gene mapping: locating the relative position of genes on a chromosome

Genetic engineering: specifically altering genes or transferring genes from one organism to another by artificial means

Haploid: a cell, such as an egg or sperm or pollen cell, that has just one complete set of chromosomes

Hemoglobin: the oxygen-carrying protein in blood, occurring in the red blood cells

Hybrid vigor: the health and strength of offspring that result from mating two parents from different inbred varieties

Inbreeding: breeding between individuals that are closely related genetically

Linkage: genes associated on the same chromosome that tend to be inherited together

Meiosis: two successive cell divisions, involving just one round of reproduction of chromosomes, that results in the formation of four haploid cells from the original diploid one

Messenger RNA: a copy of the instructions in a gene that is read in a ribosome to produce a protein molecule

Mitosis: cell division in which the number of chromosomes is kept constant, each daughter cell having exactly the same number as the parent cell

Morphogenesis: the formation of shape and pattern during development

Mutation: an inheritable change resulting from an alteration in the DNA

Natural selection: the environment favoring those organisms that are best adapted to it, and hence survive and reproduce

Polyploidy: cells or organisms with more than two complete sets of chromosomes in each cell nucleus

Protein: a large molecule comprising one or more long chains of amino acids

Recessive: a gene that does not express itself in the presence of a corresponding (dominant) partner

Ribosome: the structure in the cell cytoplasm where proteins are assembled

RNA (ribonucleic acid): a molecule similar to DNA, consisting of a single strand of bases, in which thymine is replaced by uracil (U)

Sex cells: a sperm or egg or pollen cell; the male (sperm or pollen) and female (egg) cells come together at fertilization

Sex chromosome: a chromosome, such as the X or Y, that is associated with sex determination

Sex linkage: characteristics that are associated with genes on sex chromosomes

Somatic cell: a diploid body cell, that is, not a sex cell

Somatic mutation: a mutation that occurs to a gene in a somatic cell, and that is therefore not passed on to offspring at fertilization

Species: a group of individual organisms that can breed together to produce fertile offspring

Tissue culture: growing a number of similar cells outside the whole organism in chemical solutions

Transfer RNA: small RNA molecules, each recognizing a single codon in messenger RNA in a ribosome, and inserting a particular amino acid into the growing protein chain

Virus: a minute parasite that can only reproduce inside cells of other organisms

Wild-type: the common genetic form of an organism or a particular characteristic

Index

Credits

The publishers gratefully acknowledge permission to reproduce the following illustrations:

Ardea London 43*tl*, 63, 71, 76*t*, 79; The Australian Information Service, London 80; Barnaby's Picture Library 43*tc*; BBC Hulton Picture Library 6, 87*t*; Biofotos 27, 85*t*; Biophoto Associates 11, 14, 24, 31, 53*r*, 68, 76*b*; B. Ben Bohlool 93; M. Bownes 8; Charles C. Brinton Jr 54; Camera Press Ltd 18; Carnegie Institution of Washington 45; Carolina Biological Supply Co. 26; J. Allan Cash Ltd, 81*b*; Bruce Coleman Ltd 70, 73*t*, 81*t*; Colorific 19; Daily Telegraph Colour Library 89; P. Farnsworth 64; H. Harris 85*b*; D. A. Hopwood 91; Akhtar Hussein 75; Jacana 7*b*, 9, 56, 73*b*; James Kezer, Department of Biology, University of Oregon 29; Gunther Konrad 78; L. A. MacHattie 16; The Mansell Collection 41*b*; Giuseppe Mazza 60; Ken Moreman 47*c*, *r*, 59*t*; National Heart, Lung and Blood Institute, National Institutes of Health 50; *Nature* 3; Shalom Nedan 83; Lennart Nilsson, from *Se Människan*, Bonnier Fakta Bokförlag AB, Sweden (USA and Canada: *Behold Man*, Little, Brown & Co; UK: *Behold Man*, Harrap Ltd; France: *L'Homme De Plus Près*, Jean Jacques Pauvert; Italy: *Questo E L'Uomo*, Edizioni Paoline) 48; Oxford Scientific Films Ltd 66*l*, 72; *Paris Match* 22; Claus Pelling 61; Pictor International 28; Picturepoint London 43*tr*; F. M. J. Pinn 66*r*; Plant Breeding Institute, Cambridge 33; Popperfoto 86; Poultry Research Centre, Edinburgh 43*b*; John Roberts 69*t*; Science Photo Library 30, 47*l*, 53*l*, *c*; Curt Stern 41*t*; Denis Summerbell 59*b*; Tierbilder Okapia 7*t*; Trewin Copplestone Books Ltd 20, 87*b*; Richard G. Vwrifel 32; M. E. Wallace, Consultant to Philip Harris Biological Ltd 58; ZEFA 69*b*.

Artwork by: Carol McCleeve, 41, 75; Colin Newman/Linden Artists 57, 77; all others by Illustra Design Ltd.

Cover photograph: Fritz Goro, Life Magazine/

Bibliography

Introduction to Genetics, 3rd edition, D. G. Mackean; Transatlantic, 1978

Evolution, C. Patterson; Cornell University Press, 1978

Human Genetics: An Introduction to the Principles of Heredity, S. Singer; W. H. Freeman, 1978

The Cell, 4th edition, C. P. Swanson and P. L. Webster; Prentice-Hall, 1977

Life on Earth, David Attenborough; Little, Brown & Co, 1980